Science and Social Inequality

RACE AND GENDER IN SCIENCE STUDIES

Series Editors
Evelynn Maxine Hammonds, Harvard University
Anne Fausto-Sterling, Brown University

*A list of books in the series
appears at the end of this book.*

Science and Social Inequality

Feminist and Postcolonial Issues

SANDRA HARDING

UNIVERSITY OF ILLINOIS PRESS

Urbana and Chicago

1 2 3 4 5 C P 5 4 3 2

Library of Congress Cataloging-in-Publication Data
Harding, Sandra.
Science and social inequality : feminist and postcolonial issues
/ Sandra Harding.
p.   cm. — (Race and gender in science studies)
Includes bibliographical references and index.
ISBN-13: 978-0-252-03060-4 (isbn 13 - cloth : alk. paper)
ISBN-10: 0-252-03060-5 (isbn 10 - cloth : alk. paper)
ISBN-13: 978-0-252-07304-5 (isbn 13 - paper : alk. paper)
ISBN-10: 0-252-07304-5 (isbn 10 - paper : alk. paper)
1. Progress. 2. Feminism and science. 3. Postcolonialism. 4. Science
and civilization. 5. Science—Social aspects—Developing countries.
I. Title. II. Series.
HM891.H37   2006
303.48′3′08—dc22   2005012422

To little Eva,
whose baby smiles and wiles
entertained and nourished her grandmother
throughout the final rewrite
of these essays.

Contents

Preface

These essays are intended to draw attention to a situation that has been difficult for people in the educated classes in the West to recognize but has been widely observed and discussed in social justice movements around the world. We in these educated classes, and especially university-based students, researchers, and scholars, are a significant part of the group that has most benefited from the status that scientificity can bestow. Moreover, it is we who work in science, social science, and humanities disciplines who provide the conceptual resources, often inadvertently, through which our governments and corporations can justify their disproportionate command of material resources and social control on a global scale. Through our disciplinary research and teaching, we end up servicing the "conceptual practices of power," in sociologist Dorothy Smith's phrase. We are (we think) the model of rational thinkers. Yet skeptics have been pointing out that although the benefits of modern Western sciences have disproportionately been distributed to people like us, their costs have disproportionately been borne by the economically and politically most vulnerable groups around the world. And they point out that we, who produce systematic empirical knowledge and the systemic frameworks it requires, are in denial about this situation.

The chapters to follow examine critically various aspects of this situation at this particular moment in history. My intention is to stimulate discussions of how better to harness modern Western sciences for social justice projects, not to provide definitive answers to the questions I and others raise about modern Western sciences' far too fragile links to democratic transformation. Several of the topics here are ones that I am still mulling over and trying to understand better years after I first addressed them. The world within which those issues exist has changed significantly since I first addressed them, and recent research and scholarship have also challenged familiar ways of thinking about them. I am especially interested in helping to produce epistemologies and philosophies of science that can contribute to democratic transformation; otherwise, these philosophies of science and knowledge remain part of the problem for social justice movements.

Several of the essays, or discussion of themes in them, have appeared in earlier versions. Chapter 1, "Thinking about Race and Science," did so, under the title "Science, Race, Culture, Empire," in *The Blackwell's Companion to Race and Ethnic Studies*, edited by David Goldberg and John Solomos (London: Blackwell's, 2003). Chapters 2 and 3, "Seeing Ourselves as Others See Us: Postcolonial Science Studies" and "With Both Eyes Open: A World of Sciences," draw on themes I first addressed in "Is Science Multicultural? Challenges, Resources, Opportunities, Uncertainties," in *Configurations* 2.2 (1994): 301–30, and later pursued in *Is Science Multicultural? Postcolonialisms, Feminisms, and Epistemologies* (Bloomington: Indiana University Press, 1998), though the essays here were mostly written for this volume. Chapter 4, "Northern Feminist Science Studies: New Challenges and Opportunities," is a revision and expansion of "Science and Technology," in *The Blackwell's Companion to Gender Studies*, edited by Philomena Essed, David Goldberg, and Audrey Kobayashi (London: Blackwell's, 2004). The discussion of standpoint theory in the first part of chapter 5 draws on many similar such accounts I have given over the past two decades. Chapter 6, "Feminist Science and Technology Studies at the Periphery of the Enlightenment," revises and expands the first half of "Gender, Development, and Post-Enlightenment Philosophies of Science," in *Decentering the Center: Philosophy for a Multicultural, Postcolonial, and Feminist World*, edited by Uma Narayan and Sandra Harding (Bloomington: Indiana University Press, 2000). Themes of chapter 7, "The Political Unconscious of Modern Science," were addressed in "Should Philosophies of Science Encode Democratic Ideals?" in *Science, Technology, and Democracy*, edited by Daniel Lee Kleinman (Albany: SUNY Press, 2000). Chapter 8, "Are Truth Claims in Science Dysfunctional?" is revised from an earlier version that appeared in *Philosophy of Language: The Big Questions*, edited by Andrea Nye (New York: Routledge, 1998). Chapter 9, "Does the Threat of Relativism Deserve a Panic?" is a revised version of "After Objectivism vs. Relativism," in *Toward a Feminist Philosophy of Economics*, edited by Drucilla Barker and Edith Kuiper (New York: Routledge, 2000). In fact, readers already anxious about the philosophical implications of bringing feminist, antiracist, and postcolonial studies to bear on what are supposed to be value-neutral Western sciences, epistemologies, and philosophies of science might want to begin with this last chapter, as it is designed to help allay precisely such anxieties.

Acknowledgments

I am indebted to an "invisible college" of feminist science studies scholars and researchers as well as scholars working in other fields whose work and conversations continue to inform and inspire my own thinking. This group includes Karen Barad, Cynthia Enloe, Anne Fausto-Sterling, Donna Haraway, Nancy Hartsock, Doug Kellner, Daniel Lee Kleinman, Gail Kligman, Hugh Lacey, Hilary Rose, Joseph Rouse, Joni Seager, Dorothy Smith, Banu Subramaniam, Ann Tickner, Sharon Traweek, Lisa Weasel, and Alison Wylie.

The Graduate School of Education and Information Studies at the University of California at Los Angeles has provided a welcome haven for my teaching and university participation, as well as students and colleagues whose deep commitment to social justice projects stimulates my own work.

Comments by Hilary Rose, Anne Fausto-Sterling, and an anonymous reviewer for the University of Illinois Press greatly improved this manuscript.

Finally, Dorian Harding-Morick, Emily Harding-Morick, and my darling grandbabe, Eva, along with dear friends in Los Angeles and elsewhere, have made daily life a pleasure that nourishes my soul and my keyboard practices.

Science and Social Inequality

Science and Inequality: Controversial Issues

Do modern sciences advance social progress? How one answers this question depends upon what one counts as science and as social progress. Until recently, most educated people in the West considered it irrational, a case of "science bashing," to doubt that advancing the growth of modern Western sciences contributed to social progress. Most people assumed that the term *science* does and should refer only to the West's knowledge production. More of this science could only advance social progress. Many people still hold this view. Scientific method is unique to modern Western science, they hold. The use of this method alone provides resources against superstitions and false beliefs and so ensures a contribution to social progress. Modern sciences make possible applications and technologies that raise the standard of living, improving sanitation, medicine, and health care; the ease and speed of travel and communication; and manufacturing, agriculture, and other forms of economic production, to mention only a few such benefits.

Yet for more than four decades now, skeptical views of the relation between the advance of modern Western sciences and social progress have emerged from many quarters around the globe, including from Western scientists themselves. The skeptics have focused especially on modern sciences' ties to militarism; on corporate profiteering through appeals to scientific and technological progress (the pharmaceutical industry is just the most egregious such offender); on environmental destruction caused by scientific farming, manufacturing, militarism, and many technologies in daily use; on scientific theories purportedly demonstrating racial or sexual inferiority; and on the failure of the transfer of Western sciences and technologies to the developing world to benefit the vast majority of the globe's least-advantaged citizens. To be sure, greedy and antidemocratic politics have played a role in each of these horrors. Yet the skeptics have claimed that such phenomena seem to be encouraged, enabled, and perhaps even directed by features of modern sciences themselves, not just by greed and politics that are external to them. The skeptics state or imply that more modern Western scientific research seems inevitably to increase local and global inequality and social injustice.

Scholars have joined social activists in debating just how it is that what has been widely regarded as good science, done by well-intentioned scientists, can nevertheless encourage, enable, and direct such bad effects. Historians, sociologists, ethnographers, philosophers, and cultural studies scholars have identified complicitous relations between modern sciences and their cultures' antidemocratic social, political, and economic projects.[1] Feminists, antiracists, poor people's movements, multiculturalists, postcolonialists, environmentalists, antimilitarists, Queer activists, and members of other recent prodemocratic social movements have brought into focus additional problematic aspects of relations between sciences and their societies that were heretofore undetected.

Yet to many scientists and nonscientists alike, the grounds for such dissent remain unconvincing and frequently unclear. The scientists insist that they intend to produce contributions to social progress, and that their research methods are themselves value neutral. Thus, nothing that they do as scientists could be responsible for blocking social progress. From this perspective, if an individual's intentions are good and one follows established research rules, then only good results can ensue. Of course, this cannot possibly be the case, as a moment's reflection will indicate. Yet defenders of this view of science continue to argue that when bad things happen that can seem to be caused by scientific practices, in fact these were either unpredictable or were the consequence of social and political decisions and processes that lay outside of science proper. The cognitive, technical core of science—its methods, ontology, and laws of nature—is culture free and enables modern Western sciences alone to constitute truly international sciences.[2]

Certainly, these defenders are right that both modern sciences and social relations are so complex that the social effects of scientific and technological change can be unpredictable to even the best-informed observers. Moreover, social and political processes do result in alarming scientific and technological policies that damage humans and their environments. And of course, the scientific community itself is actively engaged in exposing and exiling the perpetrators of occasional cases of "bad science."

Yet the critics focus not on these kinds of cases, but rather on the cognitive, technical core of sciences and on sciences' routine, recommended practices—everyday science that is regarded as good science. One could say that it is the very banality of modern sciences' contributions to social injustice that is at issue.[3] For the critics, it is precisely the point that scientists intend that their work produce social goods, yet it provides just the resources that already powerful groups need to increase their social advantage. The plot thickens, however. Most of these critics do want "more science." But they want ones that contribute to democratic social transformation, and these are ones that can ask the kinds of questions and produce the kinds of knowledge that the social justice movements need and want.

Clearly, different conceptions of the character of modern sciences and their participation in their cultures' moral and political projects, as well as of what should count as "social goods," are at play in such controversies. Also at issue are such other central notions as nature, scientific method, scientific rationality, value neutrality, and objectivity.

Controversial Issues

Several thorny issues, each generating disturbing questions, require further exploration to permit more fruitful dialogue on such topics among even those with shared activist or shared scholarly interests or both. Within prodemocratic movements and scholarly disciplines as well as between them, proposals for just what is wrong with contemporary scientific practices and how to improve them always stimulate controversy. Within both science activism and science studies almost as divergent views on these issues exist as among those outside such fields. The often confusing positions among participants in the recent "science wars" provide one display of such diversity and the intense emotions such issues call forth.[4] The chapters that follow focus on ways to move past these controversies to more fruitful ones. Modern Western sciences are too powerful a social force for us to want them uncontroversial. What we can reasonably want, however, is controversies that help us to think in new and fruitful ways about present and possible future relations between knowledge and power.

DO SCIENCES AND THEIR PHILOSOPHIES HAVE A "POLITICAL UNCONSCIOUS"?

Cultural critic Fredric Jameson points to the "political unconscious" of literature—to the narrative structures through which novels, for example, make unannounced political appeals, recruiting unsuspecting readers into particular social agendas. Citing Marx's argument that persistent, uninterrupted class struggles characterize all of human social history, Jameson proposes, "It is in detecting the traces of that uninterrupted narrative, in restoring to the surface of the text the repressed and buried reality of this fundamental history, that the doctrine of a political unconscious finds its function and its necessity."[5] Do modern Western sciences and their philosophies similarly have a "political unconscious," recruiting scientists and the general public into advancing local and global inequalities—of class as well as of gender, race and ethnicity, and other kinds—even as they profess to provide only culturally neutral facts through objective methods? If so, what are its distinctive features?

"Real sciences" are supposed to be transparent to the world they represent, to be value neutral. They are supposed to add no political, social, or cultural features to the representations of the world they produce, and to leave the world they observe unchanged by their research projects. Nature alone, with the assistance of universally valid rational thought and culture-free research methods, is supposed to be the sole contributor to sciences' representations of nature's order. Modern Western sciences are not supposed to have any "unconscious" at all; they are not supposed to have any conceptual practices or narrative structures unacknowledged by scientists that add cultural elements to sciences' representations of nature's order. They certainly are not supposed to have politically charged conceptual frameworks or narrative structures. The philosophies of modern Western sciences were formulated and subsequently reformulated through five centuries of social change in diverse European and later American societies, with contributions from scientists from other cultures working in Western contexts. To defenders of sciences' social neutrality, this diversity of contributors to modern Western sciences over such diversity of historical eras is in itself evidence that such sciences transcend local cultures and political projects, and that they alone are capable of doing so. They are uniquely international sciences. Yet critics point out that themes of male supremacy, racism, class exploitation, and colonial and imperial exploitation and domination, transformed from era to era and place to place, still persist throughout much of this social variation and change, as the first section of the book will show.

This situation raises troubling and hard-to-answer questions. To what extent do these inegalitarian political projects remain external to fundamentally value-free scientific assumptions, methods, and claims, and to what extent do they permeate their cognitive, technical cores, creating a "political unconscious" for sciences and their philosophies that will vary from culture to culture? If political and social values and interests do permeate deeply into sciences and their philosophies, must scientists gain skills in political philosophy and textual interpretation in order to detect and evaluate how such commitments shape what sciences count as good method and as relevant evidence for or against a hypothesis? Must philosophers do so to reevaluate standards for the objectivity and rationality of sciences? Should scientists produce such "sciences of science," or should science students be required to learn such theories and methods from the sociologists, historians, ethnographers, and philosophers of science who have developed them?

The language of egalitarian and inegalitarian, of social justice, and of social progress assumes some kind of democratic ethic. What kind of democratic ethic has in the past informed the practices of the sciences and their philosophies? What are the political visions that we should want to inform sciences and their philosophies? And who is the "we" who should make such a decision? Such issues arise throughout these essays.

WHICH PHILOSOPHIES OF SCIENCE ARE ADEQUATE FOR A WORLD OF SCIENCES?

Until recently, most people of European descent assumed that there was only one collection of beliefs and practices deserving the name "science"—modern Western ones. These beliefs and practices did already, or in principle could, form a coherent, unified representation of nature's order, or at least one that exhibited a harmonious relation among physics, chemistry, biology, and other sciences. The knowledge systems of other cultures, it was routinely asserted, were infused with magic, superstition, religion, and other forms of irrationalism and anthropomorphism, making them unreliable guides to nature's regularities and their underlying causal tendencies, and leaving the thought of those cultures firmly lodged in the premodern. Such knowledge systems did not deserve the name "sciences," and because of their cultural elements they could not be integrated into a unified or harmonious relation with modern Western sciences. This is still the prevailing view in perhaps most parts of Western sciences and in popular Western opinion. Edward O. Wilson's *Consilience: The Unity of Knowledge* has met a widespread and enthusiastic audience. The unity view is also advanced by some science intellectuals from the Third World.[6]

Certainly, modern Western sciences are in some respects far more advanced, developed, and resourceful than are the knowledge systems of other cultures. Yet the idea that it is useful to continue to conceptualize the sciences as unitary, or singular—that there is really only one science—is receiving increasingly skeptical scrutiny, including from influential tendencies in Western science studies. Indeed, considering only modern Western sciences, one school of contemporary historians and philosophers of science thinks it more useful to consider these in themselves as necessarily disunified, collectively sharing no common ontology or method.[7] So the fact that other cultures' knowledge systems have diverse ontologies and methods, different from those of Western sciences, no longer differentiates these other knowledge systems from Western sciences in themselves.

Moreover, multicultural and postcolonial science and technology observers have pointed out that other cultures' knowledge systems also have great strengths. Their achievements have frequently been adopted into modern Western sciences, and they still have valuable bodies of knowledge and inquiry techniques unavailable in modern Western sciences. In some cases, these other practices should reasonably be regarded as scientifically preferable to those of modern Western sciences. The adoption of acupuncture and indigenous pharmaceutical practices into Western medicine indicates that the Western scientific community itself understands the value of such traditional sciences. In other cultures as in the West, culture and politics have themselves been productive of scientific knowledge, not just an obstacle to it, as has also been the case. Furthermore, the cognitive diversity that different culturally embedded sciences can provide is itself a resource for everyone's scientific and technological future. Finally, a far more realistic understanding of modern Western sciences and their Eurocentric tendencies emerges if one starts off thinking about the latter from the standpoint of the history of encounters and interactions between Western sciences and other cultures' knowledge systems.[8]

It is hard to keep in focus simultaneously both "other sciences" and modern Western ones. Even referring to them as "sciences" enters widespread and vigorous controversies over what the nature of science is. Other knowledge systems seem outside the realm of the intelligible, and thus are unrecognizable as sciences, from the perspective of modern Western sciences and their philosophies.[9] What would it mean to gain such a focus in a way that avoided Eurocentrism? There are disadvantages in referring to other knowledge systems as sciences, but what are the advantages that accrue from such a practice? What is at stake, and who should get to decide, in disputes over what counts as science? Finally, what are effective strategies to avoid supporting a "tolerant pluralism" that leaves unchallenged the existing hegemonous global political economy? What are strategies to avoid merely substituting an insistence on particularism for the earlier insistence on universalism, and thereby abandoning the project of more accurately and usefully identifying shared conditions of human life and thought?

NATURE AND CULTURE

Central to debates over "the nature of science" is the question of just how far culture extends into what used to be thought of as nature's proper domain. This is an issue about the natural world that is the object of scientific study. For example, how much of racial and sexual differences can be attributed to nature and how much to culture? Yet it is also an issue about scientific practices. These are the object of study in the field of science studies—the history, sociology, ethnography, and philosophy of sciences. Both issues are addressed from diverse perspectives in the chapters that follow.

Deciding which features of humans and the world around us are a consequence of nature and which are created by culture has been a controversial issue in the West since the emergence of modern sciences. Of course, the issue has far older origins, but modern sciences seem to provide more precise and empirically compelling evidence of just where the boundary between nature and culture is to be found. To be sure, it has remained unsettling to see that other cultures have defined similar binaries differently—for example, as the raw and the cooked or the domesticated and the wild. These contrasting ontologies are by no means restricted to premodern knowledge systems; rather, they have shaped modern scientific research in the very same fields in which Western scientists study nature.[10]

For some time now, however, influential biologists and science theorists have been arguing, first, that it is misleading to assume that any characteristics of humans and our environments can be assigned to nature or culture exclusively. In the second place, a few are proposing that we need a radically different conceptual framework for biology, one that leaves no space for the misleading nature versus culture binary. This issue will arise in feminist science studies and studies of race and science. But it also arises in reflections on philosophy of science conceptual frameworks that assume that the natural sciences are capable of focusing on only what nature contributes to the world around us.

THE NEW ORGANIZATION OF SCIENTIFIC INQUIRY

How is Western scientific knowledge actually produced today? To read science textbooks or popular scientific accounts one could be forgiven for assuming that it is still heroic individual scientists and their support teams, dedicated to the search for nature's truths, who control the directions and achievements of scientific research. At the same time, however, it is clear that graduates in the natural sciences increasingly can find employment only in corporate labs; mostly, they are working for defense contractors, for pharmaceutical companies, or in electronics or biotech industries. And university science departments, which historically isolated themselves from commercial interests and now and then from national state interests, today can claim little such autonomy. Their values are commercial and national state values.

Conventional accounts of science present it as the discovery and testing of hypotheses, implying that the laws of nature had been there all along, untouched by human hands or thought, until some clever or lucky scientist managed to detect them. The conventional account emerged mainly from scientists' own narratives of their experiences, or from the historians and philosophers of science initially trained in physics or other natural sciences, for whom such accounts had stimulated them to pursue their own science training. From the scientist's point of view, the discovery account can seem like a maximally accurate description of research. To challenge its accuracy is to challenge the basis of scientific research itself, that is, the observations and recollections of the community of scientists. And it is to challenge the authority of what is perhaps perceived to be the most legitimately authoritative institution in modern societies.

Although there certainly are valuable insights in such a way of thinking about the history and practice of science, sociologists of science drew attention to how scientific knowledge is "manufactured" under constraints much like the production of a culture's other artifacts.[11] Production processes always leave their marks on their final products, and the production processes of science are no exception. Thus, distinctive concerns of particular nations, of imperial and colonial projects, and class, racial, and gender concerns all have left their marks on work in the history of science. As Thomas Kuhn famously put the point, moments in the history of science have an "integrity" with their social era.[12] He did not have in mind the gender, racial, class, colonial, and imperial aspects of historical eras, but feminist and postcolonial science and technology studies have examined what distinctive kinds of knowledge are manufactured in these contexts, too.

These sociologies of science have remained controversial for those still attached to the older "discovery" narrative. Meanwhile, social justice movements have been recommending other ways of producing the kinds of knowledge that their constituents need. Sometimes they still talk in terms of discovering aspects of nature and social relations that have been obscured by oppressive class, racial, imperial, or gender interests. But at other times they talk about how prodemocratic values and interests can also construct or constitute both the results of scientific research and the "natural" and social features of the world itself.

Scientific research is dynamic, changing over time and in response to the kinds of changes found elsewhere in a society's economic and political processes. Contradictory tendencies appear in some widely discussed accounts of recent changes in research processes. Since the emergence of the Internet three decades ago, the production and management of information have moved to the "base" of the global political economy.[13] This has had effects on how modern Western scientific knowledge is produced.[14] Furthermore, as risk theorists such as Ulrich Beck and Anthony Giddens have argued, the particular kinds of production processes responsible for modern sciences' greatest achievements are increasingly producing huge risks to human life and health, and, indeed, to the survival of the planet at all. The "risk society," in which we increasingly feel we live, is a consequence of the successes of the way knowledge has been produced by the sciences of industrial society. Apparently, the more successful modern science is, the more endangered are our lives.[15]

These social theorists join antiracist, postcolonial, and feminist scholars, as well as some of the post-Kuhnian science studies theorists, to argue that, as Beck puts the point, the production of scientific knowledge has escaped the monopoly of scientists in two ways in recent decades. On the one hand, science studies applies scientific methods to the study of scientific work itself. It expands the production of scientific knowledge *about nature* as well as about the social relations of science to the reflexive history, sociology, ethnography, and philosophy of science. On the other hand, we consumers of scientific information increasingly have to deal with making choices about our health and well-being in the face of conflicting expert opinions—for example, about safe food supplies, the safety and effect of vitamins and other health remedies, or our environments. We are forced to educate ourselves far more than our parents or grandparents felt necessary. One could say that we feel forced to join the teams of experts to produce usable scientific knowledge.

There is a third way in which the monopoly of the legitimate production of scientific knowledge has escaped its traditional producers: increasingly, non-Western knowledge systems are being recognized as fully scientific in important ways that have traditionally been attributed only to dominant Western knowledge systems. In all three ways, standards for "scientific literacy" are undergoing strengthening, expansion, and, of course, controversial debates.

SCIENTIFIC ACCOUNTABILITY AND RESPONSIBILITY:
FROM SCIENCE AS REPRESENTATION TO SCIENCE AS PRACTICE

Scientific accountability and responsibility have been at issue in all of the preceding foci. One more aspect emerges in recent attempts to reconceptualize science as a system of practices rather than as a coherent network of representations of nature or social relations or both. Conventionally, science has been thought of as fundamentally a set of statements or sentences—the laws of nature, observation sentences, and the like. Yet this way of conceptualizing it obscures how social and political values and interests seem to flow out of scientific work "behind the backs" of the scientists. The representationalist account seems to absolve the scientific enterprise of any responsibility for the various politics that flow from its representations. If no scientists intentionally put any politics into their work, then none can be assigned responsibility for any politics that emerge from this work.

Feminist work has been particularly helpful here in two ways. On the one hand, several philosophers have identified and developed accounts of just how nature always already appears to scientific observation as normative, as discursively constituted through such scientific practices as, for example, measurement or modeling. On the other hand, standpoint theory's project can be understood as identifying discursive practices that can counter the dominant ones organizing research disciplines. These two kinds of work contribute to understanding and changing the antidemocratic aspects of sciences' political unconscious.

These kinds of new work will not be uncontroversial. But it is to be hoped that the controversies that they spark can be more fruitful than some of the tired old ones that have worn out their ability to stimulate illuminating discussion and research. In several chapters I try to show why it is time for us to tuck some of these older ones into their final resting places.

Some Terminological Challenges

Some readers will already feel disoriented by the way I have been using the term *science*. Here, following usage in multicultural and postcolonial science studies, as well as in the main tendencies in post-Kuhnian history, sociology, and ethnography of science, I use it to mean any systematic empirical study of ourselves and the world around us. Thus, a knowledge system or set of inquiry practices will be referred to as a science if it is systematic and empirical, regardless of whether it is Western or non-Western, contemporary or ancient, obviously embedded in religious or other cultural beliefs or not apparently so.

Of course, there are important differences between modern physics or biology and premodern knowledge systems about nature's order. To regard all of these different knowledge (or practice) systems as sciences is not to regard them as equally accurate, comprehensive, or useful with respect to any particular questions we might ask. Some cultures understand the dangers to health of smoking; others do not. Some understand how to manage chronic pain; modern biomedicine has not until recently. Some can navigate effectively by the stars from Samoa to Australia; others would be totally lost if their global positioning system broke down. Some can get people to the moon; others cannot. Some can accurately and comprehensively grasp how cultural projects infuse their research practices; others cannot.

The benefits of conceptualizing all of these effective knowledge systems as sciences are many. One is that in "levelling the playing field" by keeping "both eyes open" when we compare modern Western sciences and other culture's knowledge systems we gain a more accurate and comprehensive grasp of the strengths and limitations of each.[16] Another is that our imaginations are invited to explore how our own knowledge systems and their relations to the economic, political, social, and cultural projects in our societies could be different than they are.

Yet one more problem clouds this rosy picture: the term itself is a Western one and can seem to commit yet one more case of Eurocentrism when I insist on using it to characterize knowledge systems not so conceptualized by their makers. I ask readers simply to remember this limitation of these essays. For the purposes of my arguments here, it is important to conceptualize on a single map humanity's attempts to understand and effectively interact with themselves and the environments they encounter. This is not the only useful map for thinking about sciences and their societies, but it is indeed a valuable one.

Another issue to which I have not found a satisfactory solution is how to refer to what used to be called the First World and the Third World. That terminology is a product of the cold war. It is problematic to import the cold war politics into our discussions here. Yet progressive science movements in the developing world, such as the Third World Network, have named themselves with such a term.[17] All of the proposed substitutes for this terminology have their own problems. "Developed" versus "underdeveloped" or "developing" countries fails to question the imperial rationale of modernization embedded in the language of development.[18] "West" versus what? The East? The Orient? The Rest? Such binaries seem clearly linked to histories of European-U.S. orientalism.[19] The language of "North" versus "South" emerged in 1992 at the United Nations' Rio de Janeiro environmental conference. In the United

States, one must say "Global North" and "Global South" to draw attention away from the histories and differences of distinctive regions within the United States. Yet even so, one cannot use the equator to divide the world since there are industrialized countries, such as Japan, in what a mapmaker would regard as the Southern Hemisphere, and many not fully industrialized ones, in Asia, the Middle East, and eastern Europe, for example, north of the equator. "Industrialized" and "unindustrialized" will distinguish cultures only from the late eighteenth century on, whereas histories of science and of European expansion, often topics in the essays that follow, are about processes that began in earlier periods.

And, of course, all of these pairs of terms force heterogeneous and complex worlds into conceptual binaries, implying that each is homogenous and undifferentiated in itself, as well as completely separated from its other. Binary making, we can remember, has been a persistent practice of the imperializing modern West. Yet we should want to reassess the qualities and practices of the cultures and their knowledge systems that the imperializing West helped to bring into existence and in many ways still supports. It is too convenient a move for the now confronted dominant group to say, "Right! We're all just humans. There are really no longer any significant differences between us as historical groups. Let's just talk about what we all share." This would obviously be absurd, though many resistant Northerners seem willing to commit such an absurdity.

Consequently, I shall shift through these various binaries as I go along, using the terms that seem most appropriate to the particular context of discussion. I shall favor "Global North" and "Global South" when referring to the contemporary world, but there are contexts, especially when referring to events in the past, where one of the other set of terms makes better sense. I ask readers to remember the dangers of reifying any binaries at all, and especially any particular one they run across here. It is worth reflecting that we should expect these kinds of terminological challenges when referring to human social practices that are themselves undergoing deep social transformations. Attempts to decrease inequalities in our "global village," as the phrase goes, require more competent terms and ways of linking them than those we have inherited.

THE IMPORTANCE OF CONTROVERSY ABOUT
SCIENCE AND SOCIETY

The essays that follow will explore these issues in a variety of ways. But they can hardly settle them once and for all—or even for the moment. Rather, they are intended to juxtapose literatures usually kept apart in order to provide more resources for thinking about what turn out to be some of the most provocative issues of our era. Conventional philosophers of science as well as the postpositivist science studies scholars rarely venture into the multicultural and postcolonial science and technology studies literature, or even into the feminist science studies literature produced by colleagues in their own departments or colleges. None of these groups have paid much attention to the sociological studies of how the Internet is reorganizing significant sectors of our societies, including the production of knowledge. Nor have the sociologists examined the relations between the social justice movements' interests in better sciences and the opportunities and effects of the Internet.

Modern Western sciences themselves emerged as part of a massive and lengthy process of shifting from the social formation centered on feudalism to one centered on Liberal democratic and capitalist political and economic relations. These essays attempt to point to contemporary links between shifts in the production of knowledge and in political and economic relations. Those of us committed to advancing social justice need public discussion of how our worlds actually function, contrary to how we are supposed to imagine that they work, and of the best strategies for distributing more justly the benefits and costs of gaining this knowledge. We appear to be in the middle of another shift in the dominant social formation, a shift that creates opportunities for the establishment of either utopian or dystopian realities.

The media today often seem to have lost sight of the nature and value of constructive controversy, substituting for it destructive criticism and the demonization of those who think differently from conventional myths. My hope is to promote the kind of stimulating and constructive controversy that can nourish the yearnings so visible around us for a better world.

PART 1

The Social World
of Scientific Research

1

Thinking about Race
and Science

Is Science Racist?

It has been tempting to ask "Is science racist?" when noticing the complicity of modern Western scientific assumptions and practices with their cultures' racist projects. Yet this may be the wrong question to ask. Too often the charge of racism has been understood as imputing intentional bias or prejudice to biologists and biomedical scientists in the personnel practices of the sciences, in the results of scientific research, and in the uses of the sciences and the technologies they make possible.

To be sure, there have been far too many such prejudiced scientists, and these kinds of intentionally racist projects certainly persist today. Yet a far more insidious and damaging project can get ignored when our focus is restricted to racial biases and prejudices.[1] It turns out that the work of many biologists and biomedical scientists has made important contributions to advancing their cultures' racist projects even when the scientists themselves have not intended such consequences of their work, and sometimes even when they have explicitly intended to recruit science for antiracist projects. Nor is it only these biological and biomedical sciences that have participated in white supremacist projects. Geography and environmental sciences have also contributed to them. Moreover, the agendas of other fields—physics, chemistry, and engineering, for example—often have prioritized scientific issues of little interest to racial minorities and largely benefiting already advantaged whites, siphoning off public funds for such projects as the U.S. space program, which are intended to demonstrate the legitimacy and desirability of global dominance by white supremacist Western societies. These are hard words for well-intentioned white Americans to hear. Yet they are frequently voiced by disadvantaged races in the United States and elsewhere in the West, not to mention by former colonial subjects around the world—a topic to be explored in the next two chapters.

We would do better to ask a different question: under what conditions could it occur that a society with widespread and powerful forms of structural racism—a race-segregated social structure—could produce sciences that did *not* participate in justifying and maintaining such white supremacy?[2] The problems of interest to a culture's sciences at a given moment in history, the hypotheses proposed to explain such problems, the methods chosen to test those hypotheses, decisions about what should count as evidence for or against such hypotheses, or the goals to be achieved in resolving the problem—how could these aspects of sciences *not* contribute to maintaining the existing social structure and agendas of the white supremacist society that decides which scientific projects to support? This question directs us to a richer understanding of the power, or lack of it, of scientists' intentions. A white supremacist society need not be one in which all or any white individuals intend or prefer their supremacy. It can also reasonably designate societies where most whites report that they oppose white supremacy, yet the values and social structures of the society de facto maintain racial inequality. In such cases, scientists can end up advancing white supremacist agendas though they have no intention of doing so.[3] And this second point directs us toward identifying strategies to counter the unintended complicity of sciences with their cultures' white supremacist agendas.

Of course, many scientists have actively struggled against white supremacy and do so today. They have devoted their scientific careers to demonstrating the falsity of popular racist assumptions and of results of scientific research used to justify such assumptions. These important projects have widely influenced what many white Americans and Europeans think about racial differences. Yet for most of the history of modern racism, white supremacist agendas and practices, in the sciences and outside them, have been openly advocated. The spread of ideals of liberal "toleration" has made the public expression of forms of bias and discrimination inappropriate. But much white supremacist belief and practice can continue without public expression or even private awareness of it; it has become largely covert or institutional.

One answer to our new question about how to counter the complicity of the sciences with white supremacist agendas is that the emergence of a powerful antiracist social movement can generate alternative science projects in such a context. Such projects can identify and protest not only the white racist agendas of sciences but also the institutional practices and philosophic assumptions that make white supremacist outcomes seem natural, banal, even to smart and well-intentioned scientists. Unfortunately, for most of the history of modern Western sciences, such movements have not existed or have not been powerful enough to access resources to make their case in public spaces controlled by the dominant race. Once the possibility (not to mention desirability, as discussed in the Introduction) of achieving culture-free sciences is abandoned, it becomes reasonable to evaluate more critically how modern Western sciences are and are not complicitous with the explicitly racist and implicitly white supremacist projects of their societies. The abandonment of white supremacist denial and self-delusion opens up the possibility of critical understandings that are otherwise blocked. Of course, racism is not only a set of conscious and unconscious beliefs and assumptions. It is most powerfully a set of economic and social practices that distribute benefits of such practices to some groups and their costs to others. So "the truth" will not set us free in this case, any more than it can do so in any other. Yet it is better, all things considered, to have an accurate understanding of how racism functions—in our case here, in sciences and their applications and technologies.

Defenses of Racist Scientific Practices

Three practices with racially discriminatory consequences were the first on which critics focused. These criticisms were rejected as misfocused by the defenders of value-neutral science. The first such practice sought to define natural racial types of humans. From the beginning of the nineteenth century until well after World War II, biological and biomedical sciences sought to divide and rank human groups, persistently "discovering" the natural inferiority of non-Europeans, as well as Jews, women, and other politically, economically, and socially vulnerable groups. Craniology was just one such now discredited comparative science; skin color, hair texture, and virtually every other body feature were carefully compared in the white supremacist projects of these sciences.[4] A second critical focus has been on racist misuses and abuses of the sciences and their applications and technologies, such as Nazi eugenics, the Tuskegee syphilis experiments, discriminatory testing and uses of reproductive technologies, and the environmental racism that disproportionately locates toxic industries and dumps in nonwhite neighborhoods and Third World societies.[5] A third critical focus has been on the ways that people of non-European descent have been disfavored in the social structure of European and U.S. science through exclusion, marginalization, and restriction to lower-level jobs.[6]

Those who reject such criticisms, whom I shall refer to as the defenders of autonomous or neutral science, often acknowledge that such projects have been unjust and ill-founded. Nevertheless, they reply, the racism of such projects does not challenge the fundamental cultural neutrality of autonomous sciences. Modern science itself cannot legitimately be charged with racism on the basis of these kinds of criticisms. In the first case of attempts to discover the natural distinctions between the races, the vast majority of such projects—such as craniology or other problematic comparative studies of intelligence or of body parts—are simply examples of bad science, not of real science, the defenders of value-neutral science say. The second and third criticisms point only to bad social practices, the defenders argue, not to racism within the sciences themselves. How the information produced by neutral sciences is used and who gets to do science are both social matters decided in civic life, not scientific matters controlled by the rigorous methods that can create culturally neutral sciences. Furthermore, these defenders of autonomous science conceptualize the racism claimed to be apparent in these practices, if it does indeed exist, to be the consequence of false beliefs and bad attitudes of unenlightened individuals. The production of accurate scientific information is the best way to eliminate the appeal of such prejudices and biases, they argue.

Yet the antiracist critics do not find compelling these defenses of the autonomy of science from society, this way of conceptualizing racism, or this remedy for racist practices. For one thing, they point out that we now have many decades of more accurate scientific information about racial differences and similarities between the races, and of the inadequacies of the very concept of racial types to explain human variation. Race is not natural in the ways assumed by studies of racial types. Moreover, many scientists who are not overtly prejudiced in the sense that they do not hold false beliefs or have "bad attitudes" toward nonwhites continue to engage in all three kinds of scientific projects that have racially discriminatory consequences. Why has the widespread availability of more accurate information about purported racial difference had so little positive effect on scientific practices?

From the very beginning of the early criticisms could be heard hints of deeper links among scientific, racial, and Eurocentric projects. Analyses of the causes of these three criticisms in terms of individuals' prejudices could not capture such links. Is racism limited to the intentional acts of individuals, as the prejudice analyses assume? What about discriminatory principles, projects, practices, and cultures of institutions such as the law, the economy, or modern scientific institutions and their philosophies? What about racist and Eurocentric assumptions made by whole societies, not just by individuals, and assumptions held over even larger historical eras and cultures, such as "modern life," "the developed world," or five centuries of European expansion? Are individuals' biases and prejudices the causes or effects of such larger social and historical eras and their projects?

Slowly but surely, a horrible truth has come to light: the smartest and best-intentioned individuals can find themselves contributing to what other cultures and later eras identify as racist, white supremacist, and Eurocentric projects. Today, much antiracism work has focused on helping individuals to improve themselves, to become antiracist individuals. They learn to identify racist biases and prejudices in everyday life, and to try to eliminate them from their own beliefs and behaviors and to help others to do so also. Such work is valuable, but it will have little effect on changing racist social structures and widely shared assumptions unless it is actively put in the service of an antiracist political movement. The self-improvement of individuals is never an adequate substitute for collective political action against the white supremacist interests, policies, and practices of dominant social institutions. Research disciplines, their conferences and publishing practices, museums and public education projects, in the social sciences, humanities, natural sciences, and arts are included in the dominant social institutions that have had white supremacist interests, policies, and practices. Thus, some of the most powerful recent analyses have sought to identify racist and ethnocentric assumptions and practices of First World institutions, societies, and civilizations, ones that are to be found beyond or outside the intentions of individuals.

The next three sections identify current issues in thinking about the natural or cultural elements (or both) of racial types, racist misuses and abuses of the sciences and their applications and technologies, and racism in the social structures of sciences. Surprisingly, the progressive ways older racist accounts have been rejected turn out to have their own problems—which, let me immediately say, is not to argue for a return to the older racist accounts. Next, this essay points to new directions in research on other cultures' science and technology traditions and on the connections among European expansion, the growth of modern sciences in Europe, and the decline of other cultures' scientific traditions, a topic to be pursued further in chapters 2 and 3. It concludes by identifying some of the challenges to conventional philosophies of science and epistemologies—those deepest and most pervasive homes for white supremacy and Eurocentrism—that have emerged from research on race and sciences.

NATURAL RACIAL TYPES?

Three lines of thinking have developed in response to the question of whether there are natural race differences and, if so, what they are. The first assumes that there is a "truth of race" that is part of the natural order, can be discovered by biology, and has valuable social implications and consequences. The second argues that race is entirely a socially constructed category, that it is not a useful scientific concept, and that these facts destroy the foundation for racial prejudice and discrimination. The third challenges both of these views, arguing that the second, like the first, insufficiently questions the nature-culture dichotomy that provides the foundation for scientific racial difference debates. This third line of thinking points out that since bodies are physiological entities that are both conceptualized in different ways in different cultures and, consequently, materially shaped by social practices, race is always both biological and social. What we need to learn is just how social practices produce socially and biologically raced bodies. This view has huge implications for the foundations of contemporary biology more generally.

The "natural races" view has a long and by now well-documented scientific history, emerging at the end of the eighteenth century and flourishing by the mid-nineteenth century.[7] These sciences employed different methods at different times, favoring morphology and classification in the eighteenth century, but shifting to comparative histology, functional analyses, and the analysis of bodies' internal organization in the nineteenth century. Measuring racial differences in intelligence is just one of the projects in this history. Scientists have sought racial classification of human differences also in the shapes of skulls, lips, noses, foreheads, pelvises, or sexual organs; in sensitivity to pain; in genetic or hormonal makeup; and in skin color, hair texture, and yet other traits. It is disturbing to discover that such projects have been pursued by some of the most distinguished scientists using state-of-the-art methods for their day. Nor were these scientists disproportionately racist; many were among the most politically progressive figures of their day on racial and gender issues. As historians have pointed out, Nazi eugenics programs were only following the lead of mainstream scientific research in the United States and elsewhere in Europe.[8] Although the emergence of population genetics in the 1930s and '40s marginalized the older focus on racial typologies, traces of the older search for biological determinants of racial differences have lingered on in such fields as the IQ debates and controversies over sociobiology.[9] Indeed, assumptions of natural races are by no means archaic, since racial classification systems remain useful in physical anthropology, forensic pathology, physiology, and some areas of public health, and as we will see, they are having a resurgence in genetic research.[10]

The second approach has rejected the "natural race" view in favor of the argument that racial categories are always socially constructed.[11] Recognition of the World War II atrocities committed in the name of maintaining racial purity decreased the attractiveness to scientists and their funders of continuing to search for biological determinants of racial differences. Moreover, the social race defenders pointed out the immense cultural variation in systems of racial classification. The United States had had a complex system of classifying people of African descent as quadroons, octoroons, and those legally counted as black because their ancestry gave them $\frac{1}{32}$ "black blood." In parts of the Caribbean and Latin America, class has shaped racial classification, so that the richer one is the whiter one is perceived to be. South Africa, Japan, and other cultures have their own systems of racial classification.

Additionally, the emergence of population genetics made attractive the practice of treating alike variation in human populations and in animal and plant populations. Thus, purported racial difference is best understood on the model of classifying butterflies or petunias. By the 1960s this social race approach was well established in biology and the social sciences, and by the 1990s the United Nations Educational, Scientific, and Cultural Organization (UNESCO) had redefined race as entirely a matter of social practices. For example, culturally established "mating" ideals shape genetic distributions, such as those that occur through limiting or expanding the numbers of mixed-race children in a community. (Consider that Spanish colonial policies sometimes encouraged and at other times forbade marriage between the conquistadores and indigenous women of the Americas. The British had similarly shifting policies in India and Africa.) State racial and class policies affect access to nutrition and health care and exposure to environmental toxins, natural hazards, and dangerous working conditions. According to many who understand race as entirely socially defined, racial prejudice, now deprived of its scientific grounds, should soon wither away.

A third approach argues that the social race position is also flawed.[12] For one thing, any search for racial differences will always be able to find them, since one can always find at least some similarities or differences between any two objects or collections of them. Thus, the apparent reality of various scientific criteria for racial difference is a product of searches for difference rather than a preexisting aspect of natural or social orders.

Moreover, it is already encultured material bodies that scientists observe. The bodies that scientists observe already have meanings and values given to them by distinctive religious and other cultural discourses, as well as by the history of Western scientific mechanist views. But scientific practices themselves also enculture bodies: scientists observe, measure, and intervene in bodies in culturally distinctive ways. From scientific practices flow cultural and political values and assumptions, "behind the backs," so to speak, of the scientists themselves. The social race view substitutes reification of the social determinants of race for the natural determinants favored in the older view. Thus, it reinstates the problematic Western assumption underlying both positions that nature and culture are discrete categories.

Most important, biological "race talk" has reemerged in highly funded genetic research. The widespread appeal of a purely social conception of racial differences has not blocked a powerful return to assertions about natural racial types. This research promises to deliver huge benefits to the researchers as well as to the pharmaceutical industry, while doing little to solve the large-scale health problems that were used to justify enormous research expenditures.[13] Hotly debated is whether this research will further stigmatize African American and other "raced" bodies, or whether it will in fact deliver some health benefits to them. However, Anne Fausto-Sterling argues, the main problem with this research

> is not what the new biology of race will produce, so much as it is that the new biology of race diverts our attention from solving problems using solutions we already have at hand. It does so in part by insisting that genes build bodies outside of culture, and then deposit them on earth, where social systems tinker with them just a bit. We need, instead, to develop the habit of thinking about genes as part of gene-environment systems, operating within networks that produce new physiologies in response to social conditions. In this view, bodies are not static slaves to their biology. Rather it is our biological nature to generate physiological responses to our environment and experience. We use genes to produce such responses.[14]

What we need to seek is not the "truth of racial difference" in nature or in social relations, but how scientific as well as social practices do change bodies in ways that conform to racial projects, as Fausto-Sterling and historian of science Evelynn Hammonds argue, as well as historian of primatology Donna Haraway.[15] Bodies are always both natural and cultural, both biologically and socially raced. We need a new biology, Fausto-Sterling argues, that is capable of thinking about gene-environment systems.

These three approaches reveal in different ways how thinking about racial difference has always been permeated by a focus on sex difference also. The critics of "natural race" have approached this topic through, for example, analyses of scientific uses of race-sex analogies, of the manipulation of reproductive practices to preserve racial and culturally valued sexual differences, and of primate studies that have been used to generate scientific arguments for the service of racial and sexual stereotypes to racist, colonial, and androcentric projects.[16]

THE RACIST MISUSE AND ABUSE OF SCIENCES AND THEIR APPLICATIONS AND TECHNOLOGIES

A second focus of criticism has been on the uses of scientific information and technologies for racist projects. Sterilizing African American welfare recipients; medical experimentation on Jews, Gypsies, African Americans, and Puerto Ricans; patterns of environmental toxicity and environmental destruction due to military and industrial practices, and the racially discriminatory uses of reproductive technologies are some of the best-known examples of such projects.[17] These criticisms, like the other two, support the more general argument that the benefits of modern sciences and technologies have been disproportionately distributed to the already most economically and politically advantaged groups and the costs to the already least-advantaged groups.[18] Moreover, critics point out that mainstream philosophies of science obscure such patterns by an accounting system that disclaims any responsibility the sciences might have for discriminatory or destructive uses or consequences of the sciences, foreseen or unforeseen, yet claims credit for any and all good uses and consequences of sciences, foreseen or not. Defenders of pure science then argue that because individual scientists often do not intend such consequences, it is simply wrong to attribute any responsibility for such bad uses and consequences to the sciences. On this improbable view, sciences are not fundamentally social institutions intricately imbedded in larger social formations. Instead, they are only a collection of value-free methods, pieces of information disconnected from any social aspects of their production, use, or consequences and the purely technical (and thus value-free) intentions of basic scientific researchers.

Of course, it would be absurd to blame the sciences alone for the problematic applications and technologies identified by the critics, let alone to blame individual scientists. That is not the point of these criticisms of the neutral science position. Rather, it is to understand the functions of scientific projects in their historical contexts, and how racist, white supremacist, and Eurocentric consequences follow from apparently innocent assumptions about human variation, causes of social change, human progress, the functions of value neutrality, and what constitute good methods. Science-and-society constitute one social formation in each society or subculture, each aspect of which is rooted in assumptions of the other. This is the theme of discussions in chapters that follow. My point here is that scientific institutions, as well as the rest of us, need to take responsibility for the ways that white supremacist consequences seem to flow from what were thought to be objective, value-neutral scientific practices.

RACIST SOCIAL STRUCTURES IN THE SCIENCES

A third concern has been with racist patterns of discrimination in the social structure of the sciences. Gaining access to scientific training and jobs has often required heroic struggles. It has required racial "minorities" to tolerate insulting and demeaning treatment by educational and scientific institutions and by racially privileged teachers, employers, and peers. Even when trained as scientists, few people of color have been promoted to the most distinguished teaching positions, directorships of the most prestigious laboratories, or the most powerful science policy positions.[19]

Various explanations have been offered for these patterns. Until the 1960s U.S. racial minorities had only restricted access to higher education (in some states, even to high school) and to professional-track preparation for a scientific career. Moreover, it has been difficult for youngsters in these groups to see themselves as scientists, mathematicians, or engineers given the absence of relevant role models. Nor have they had access to mentors—white or not— as have their white peers. Indeed, recent studies suggest that at least elementary school science teachers have such trouble envisioning racial minority girls as scientists that they persistently discourage precisely those children in such groups who would be most likely to have successful scientific careers.[20] Furthermore, a community service ethic has tended to direct career choices for racial and ethnic minorities. Given the sciences' other racist and Eurocentric theories and practices, these fields have not ranked high as enabling service to minority communities. We can note that something like a "community service ethic" overtly attracts many whites to scientific careers, since advances in science are routinely rhetorically associated with progress for "civilization," where *civilization* is defined in terms of distance from the lives of "primitive peoples."

There have been notable exceptions to such generalizations. In the early part of the twentieth century, biologist George Washington Carver experimented with peanuts in order to create products useful for African Americans that they could themselves produce. In the 1920s and '30s African American Ernest Everett Just made important contributions to developmental biology. African American Charles Drew discovered blood platelets. And in the past few decades many more U.S. scientists of color have risen to the top of scientific institutions.[21] Moreover, people of color have seized opportunities for careers in science and medicine when they have appeared. For example, Darlene Clark Hine has documented the struggles and achievements of 115 African American women who received medical degrees in the quarter century following the end of U.S. slavery.[22] Furthermore, some of the achievements attributed to scientists and engineers of European descent have been misattributed; they were produced by their slaves, servants, or other employees (as well as their sisters and wives)—a pattern found in the history of science in Europe as well.

In recent years there have been intensified efforts to increase the numbers of minorities in the sciences, maths, and engineering industries through improving elementary and high school science education and introducing outreach programs in universities and industries.[23] We are used to seeing pictures of at least token African Americans, Hispanics, and Asians in photographs of laboratories, university science classes, and ads for science, medical, and engineering products and jobs. Scientific honors now occasionally go to scientists of non-European descent in the United States. Yet there is a long way to go before racial discrimination disappears from the social structure of the sciences.

Today, a walk through any lab—university or industrial—will reveal large numbers of scientists of non-European descent. Many, perhaps even the majority, are foreign born. Becoming an engineer or scientist has always been a path to upward mobility in the United States. It is now an attractive way for people in many Third World societies to join an international elite, as well as to enjoy the other benefits that come from a career in these fields. Some Third World societies, such as India, have explicitly invested in scientific and engineering training as a way to increase their global status as well as to reap military and commercial benefits.[24] Indeed, the "brain drain" in the sciences is one way in which human resources from around the world continue to be appropriated into Northern projects half a century after the beginning of the end of formal European and U.S. colonial rule. Yet even this is an advance of sorts against earlier forms of racism that strictly limited access to scientific educations and careers to people of European descent. As historian Michael Adas points out, an extreme form of this policy was visible in the refusal of the British rulers of India to let Indian science and math students even learn of the contributions to international mathematics that had been made by distinguished Indian mathematicians.[25]

*   *   *

These are the three kinds of practices that initially led to charges that modern Western sciences are racist. Yet they occur against a larger background of Eurocentric assumptions and practices with overtones of racism and white supremacy that have been more difficult to bring into focus. We will look at some of these assumptions in the next two chapters. Here we can try to identify both the effects of such Eurocentric histories of science on racial minorities within the West and North, such as in the United States, and the resources that anti-Eurocentric projects bring to antiracist ones.

A World of Sciences: Race, Culture, and Empire

We will see in the next two chapters how recent research in comparative ethnosciences has challenged the conventional contrast between real, modern, transcultural European sciences, on the one hand, and the mere ethnosciences of other cultures, on the other hand. A culture's location in heterogeneous nature, the particular foci of its interests in surviving and flourishing, the discursive resources of its religious and cultural traditions, and its ways of organizing the labor of inquiry—all of these will contribute to its production of a distinctive body of knowledge and ways of obtaining it.[26] This is the case in the modern West no less than in other cultures. Within Western societies, too, subcultures develop their own knowledge systems. In the United States, Asian American and Native American communities have brought with them diverse medical and health traditions; poor rural communities and poor urban communities have developed skills at remaining healthy in their particular natural and social environments in the absence of medical insurance; elderly caretakers of sick kin have learned how to negotiate with and help to direct physicians' standard practices; midwives learn ways to assist in birthing that have not been taught to gynecologists; farmers, fisherman, and hunters learn to interpret patterns in nature invisible to others; thieves learn tricks of their trade one step ahead of the security systems intended to block them; and hackers learn secrets of the Internet undetected by software designers. Some of these subcultures bring with them useful parts of the knowledge system of a society of origin. Then they develop distinctive bodies of knowledge through their encounters and exchanges with the new society. Such subcultures learn, for example, when to turn to Western biomedicine and when to resort to traditional medical and pharmacological resources. Others continually work to stay one step ahead of those who would foil their uses of their knowledge. In both kinds of cases, these bodies of knowledge are empirically based and dynamic, constantly changing to meet the challenges of changing social and natural environments.

Europe's so-called voyages of discovery brought European cultures into encounters with many cultures of peoples of color. Many aspects of European sciences were developed specifically to enable the expansion of European empires and the formation of European colonies around the world at the expense of the peoples they encountered, as we will see in the next chapter. Indeed, modern forms of racism were developed precisely as parts of the orientalism that justified the conquest of Europe's racial and cultural Others.[27] It is impossible to separate racism from colonialism and imperialism and the development of modern science in Europe in this particular historical era.

The critiques of First World development policies in the Third World (see next chapter) show how it has been primarily "dedevelopment" and "maldevelopment" that the West has delivered to the Third World through its "transfer" of Northern technologies and their scientific rationality to the South. It is the investing classes in the North and their allies in the South who have been the greatest beneficiaries of development policies. Thus, the transfer of Northern science and technology to the South has increased the gap between the "haves" and the "have-nots." Within the North, a similar process is apparent in, for example, white society's recruitment of minority labor into the lower ranks of factory, hospital, laboratory, electronic, and other technical work, where their low wages and hard working conditions generate profit for those at the top. Women in both the North and the South are especially susceptible to these recruitment policies since they are already the more political and economically vulnerable gender.

Finally, epistemologies and philosophies of science have not been and cannot be immune from these kinds of racial, imperial, and colonial projects. Modern Western sciences and their philosophy have a "white supremacist unconscious."[28] The sciences and philosophies of science of peoples of European descent are presumed to be purified of the religious and cultural features of their societies, whereas those of peoples of color are considered impure, permeated through and through by their religious and cultural features. The very standards for objectivity, rationality, and good method have been constituted in terms of their distance from qualities and practices associated with "primitives." Peoples of European descent are presumed to be capable of dispassionate objectivity, rationality, and higher mental achievements. "Primitives" are claimed to be ruled by their passions and bodily needs; they tend toward subjectivity and irrationality and are incapable of higher mental achievements. Even when presented with descriptions of the Christian, explicitly Protestant, and liberal democratic aspects of modern Western sciences and their societies, or even of the specifically French versus English, or other national features, peoples of European descent cannot seem to grasp that therefore modern Western sciences have powerful culturally specific elements, as do other cultures' knowledge systems. They still deny that other cultures' knowledge systems are significant parts of the valuable repositories of human knowledge about the natural and social worlds. U.S. racial minorities bear the burden of such negative assumptions as these occur in epistemologies and philosophies of science, too.

Moreover, as theorists of science and modernity consistently point out, the underdevelopment of epistemologies and philosophies of Western sciences have socially and politically regressive consequences for everyone, but especially for those who are already politically and economically most vulnerable. The risks to life and health modern Western sciences and their technologies enable through the production of toxic environments, the increasing frequency of pandemics, and the escalating horrors of global militarism all have their worst effects on the already most vulnerable of the world's citizens, a vast majority of whom are not white.

Conclusion: The Radical Role of Antiracist Resistance

Resistance to racial features of modern sciences and their philosophies has a radical transformative potential. Of course, such resistance projects serve racially oppressed communities in valuable ways. But they also offer possibilities of expanded self-understanding to peoples of European descent. As we rethink the dominant legacy, we need to do so with both eyes open—one on the history and practices of modern Western sciences, and the other on the history and practices of the peoples of color whom modern Western cultures have encountered.[29]

2

Seeing Ourselves as Others See Us:
Postcolonial Science Studies

"Would the Gift the Genie Give Us . . ."

Maximizing objectivity requires not just that we accurately represent the way we see ourselves, others, and the world around us but also that we take seriously how others see us, themselves, and the world. Indeed, such a principle has been institutionalized in modern Western scientific practice in the directive that scientific observations must be repeatable by other observers. Others must be able to confirm what we claim were our inquiry actions and their outcomes.

Western science has usually restricted this principle, however, in two ways that we can now see have had discriminatory effects. It is only the observations of "informed" or "well-trained" observers that count. Thus, it is only the members of the scientific community, understood as the professional scientists in that community, that can count. Scientists are accountable only to each other's observations of nature and social relations, according to this view. One might wonder what could be objectionable about such an apparently reasonable restriction. After all, the scientific community should not be expected to pay serious attention to every observation or opinion offered by people outside the community, by nonscientists. There are a lot of ungrounded opinions, not to mention kooks, out there! Indeed, scientists have to make hard decisions about with which proposals and criticisms they will engage, even when their choices are restricted to ones coming from their well-informed colleagues.

Yet there can be bad effects on the sciences and their social consequences when such restricted choices are made about to whom the scientific community should listen. It turns out that both social justice and the growth of scientific knowledge are blocked when criticisms and knowledge claims of progressive social movements are ignored or dismissed. Of course, it is not that every claim made by feminists, environmentalists, poor people, antimilitarists, indigenous peoples, discriminated-against races and ethnicities, or postcolonial and other progressive groups around the world must be regarded automatically as true. Instead, the point is that it is now widely recognized that these groups have identified serious limitations in the standard ways of thinking about modern Western scientific practices and the nature that scientists study. They have themselves made important contributions to the growth of scientific knowledge. Thus, standard ways of thinking about objectivity seem excessively to restrict to whom the sciences should be accountable.

The second restriction asserts that the consequences of scientific inquiry for which science can be claimed responsible are only those intended by scientists, and as scientists their responsibility extends only to the laboratory door, so to speak. According to this standard view, science cannot be held responsible for any of its bad effects that occur outside the laboratory or after the research ends, even if these consequences are perfectly predictable by them or by others. Yet this assumption ignores the ways institutional, political, and cultural values and interests have shaped how modern Western sciences interact with the world around them and, consequently, what they "know." It encourages scientists to cultivate systematic ignorance about how scientific institutions and practices actually work and about the likely consequences of their projects.

This chapter focuses on the criticisms of these limitations that have been voiced in the multicultural and postcolonial science studies movements. It focuses on problems with familiar boundaries on which conventional sciences and their philosophies have insisted.[1] We begin with a puzzling question.

Are the Natural Sciences Multicultural?

Could they and should they be? Such questions initially may seem ignorant or, at least, odd, since it is exactly the lack of cultural fingerprints that is conventionally held responsible for the great successes of the sciences. The sciences "work," they are universally valid, it is said, because they transcend culture. They can tell us how nature really functions instead of only how the British, Native Americans, or Chinese fear or want it to work. Perhaps one can reasonably imagine that certain areas of perfectly good medical and biological sciences have been shaped by differing religious and other cultural understandings of bodies and of nature. But physics, chemistry, and genetics?

Plausible as such resistance to the idea of multicultural sciences seems, there are good reasons to doubt if one should regard it as ending the matter, however. Multicultural perspectives, often guided by postcolonial projects, are providing more comprehensive and less distorted understandings of history, literature, the arts, and social sciences. They are beginning to reshape public consciousness as they are disseminated through television specials and new elementary and high school history and literature textbooks. Indeed, daily news reports often take perspectives on the West that conflict with the conventional beliefs. Many Westerners now understand these conventional beliefs to be Eurocentric. Is it reasonable to suppose that the challenges raised by multicultural perspectives in other fields have no consequences for the natural sciences?

There are three central issues for anyone who wishes to explore this issue. First, to what extent does modern science have origins in non-European cultures and thus in an important sense already exhibit multicultural elements? Relatedly, how should we conceptualize those elements of other cultures' knowledge systems that were good enough to be adopted into modern Western sciences? Second, could there be and have there been other sciences, culturally distinctive ones, that also "work" in their particular environments and thus are universal in this sense? Third, in what ways is modern science itself distinctively European or European American, and thus culturally "local" in this sense? Fortunately, pursuit of these questions has been made easier by the appearance in English since the early 1980s of an increasingly rich set of writings on such topics. These postcolonial science studies are authored by scientists and engineers and by anthropologists and historians of science. These authors are of European and Third World descent; both groups may be found living in the South and in the North.[2]

I have noted in the introductory chapter how the terms of this discussion are and must be controversial, for who gets to name natural and social realities gets to control how they will be organized. Moreover, it is not just language that is at issue, but also a "discourse"—a conceptual framework or set of practices with its logic linking my words in ways already familiar to readers—that is adequate to the project of this essay. *Science, Western, postcolonial,* and *Eurocentrism* are terms that are linked in nontraditional ways by the arguments in the literatures of interest here. Their referents and meanings will become clear as we follow the arguments.

Let us turn to the three questions that will help to determine the degree to which science may be multicultural.

DOES MODERN SCIENCE HAVE NON-WESTERN ORIGINS?

Least controversial is to acknowledge that modern sciences have borrowed from other cultures. Most people are aware of at least a couple of such examples. However, the borrowings have been far more extensive and important for the development of modern sciences than the conventional histories reveal. Modern sciences have been enriched by contributions not only from the so-called complex cultures of China, India, and others in east Asian and Arabic societies but also from the so-called simpler ones of Africa, pre-Columbian Americas, and others that interacted with the expansion of European cultures.[3]

To list just a few examples, Egyptian mystical philosophies and premodern European alchemical traditions were far more useful to the development of sciences in Europe than is suggested by the conventional view that these are only irrational and marginally valuable elements of immature Western sciences. The Greek legacy of scientific and mathematical thought was not only fortuitously preserved but also developed in Islamic culture, to be claimed by the sciences of the European Renaissance. Furthermore, the identification of Greek culture as European is questionable on several counts. For one thing, the idea of Europe and the social relations such an idea made possible came into existence only centuries later. Some would date the emergence of Europe to Charlemagne's achievements, others to fifteenth-century events. Moreover, through the spread of Islam, diverse cultures of Africa and Asia can also claim Greek culture as their legacy.[4]

Some knowledge traditions that were appropriated and fully integrated into modern sciences are not acknowledged at all. Thus, the principles of pre-Columbian agriculture in South America, which provided potatoes for almost every European ecological niche and thereby had a powerful effect on the nutrition, health, and longevity of Europeans, and consequently on the subsequent history of Europe, were subsumed into European science; today, these origins are unacknowledged.[5] Mathematical achievements from India and Arabic cultures provide other examples. The magnetic needle, rudder, gunpowder, and many other technologies useful to Europeans and the advance of their sciences (were these not part of scientific instrumentation and thus method?) were borrowed from China. Knowledge of local geographies, geologies, animals, plants, classification schemes, medicines, pharmacologies, agriculture, navigational techniques, and local cultures that formed significant parts of European sciences' picture of nature were provided in part by the knowledge traditions of non-Europeans. ("We took on board a native of the region, and dropped him off six weeks further up the coast," report the voyagers' accounts.) Summarizing the consequences for modern sciences of British imperialism in India, one recent account points out that in effect, "India was added as a laboratory to the edifice of modern science."[6] We could say the same for all of the lands to which the "voyages of discovery" and later colonization projects took the Europeans.[7]

Thus, modern science already is multicultural at least in this sense that elements of the knowledge traditions of many different non-European cultures have been incorporated into it. There is nothing unusual about such scientific borrowing. It is evident in the ordinary, everyday borrowing that occurs when scientists revive models, metaphors, procedures, technologies, or other ideas from older European scientific traditions, when they borrow such elements from the culture outside their laboratories and field stations or from other contemporary sciences.[8] After all, a major point of professional conferences and international exchange programs, not to mention "keeping up with the literature," is to permit everyone to borrow everyone else's achievements. As we shall shortly see, without such possibilities, sciences wither and lose their creativity. What is at issue here is instead the Eurocentric failure to acknowledge the origins and importance to "real science" of these borrowings from non-European cultures, and thereby to trivialize the achievements of their scientific traditions. Such acknowledgment also raises the specter that Western claims to the one and only real science may not stand up to empirical examination. Recognition of such "borrowings" also raises issues about the often brutal and imperial methods through which they occurred.

Yet to give up this piece of Eurocentrism does not challenge the obvious accomplishments of modern sciences. Every thinking person should be able to accept the claim that modern science is multicultural in this sense. Of course, it is one thing to accept the legitimacy of a claim that conflicts with one's own and quite another to use it to transform one's own thinking. To do the latter would require that historians of science and the rest of us Europeans and European Americans locate our accounts on a global civilizational map rather than only on the Eurocentric map of Europe that we all learned.

There are implications here also for philosophies and social studies of science. For example, the standard contrasts between the objectivity, rationality, and progressiveness of modern scientific thought versus the only locally valid, irrational, anthropomorphic, and backward or primitive thought of other cultures begin to seem less accurate and explanatorily useful after even this initial stage of the postcolonial accounts. Whether overtly stated or only discreetly assumed, these contrasts damage our ability not only to appreciate the strengths of other scientific traditions but also to grasp what are the real strengths and limitations of modern sciences.

Another issue is how to conceptualize these elements of other cultures' sciences that are subsequently adopted into modern Western sciences. Are they scientific elements of prescientific thought—or perhaps even of ethnoscientific magic and superstition—until they enter Western sciences? Do inquiry elements become scientific only when Western sciences and their philosophies can recognize them as such? I return to such issues below.

These accounts of multicultural origins begin to create uneasiness about conventional beliefs that modern sciences uniquely deserve to be designated sciences, but they do not directly challenge them. They do not yet directly challenge the conventional conflation of "science" with the "Science" of European origin.[9] Nor do they challenge the assumption that the European sciences are universally valid because their cognitive or technical core transcends culture. Other arguments in the postcolonial accounts do.

Could there be, or are there already, other culturally distinctive sciences that also "work"? The postcolonial accounts have shown how rich and sophisticated were the scientific traditions of Asia, Islam, and "simpler" societies of the past. But what about the present and future?

COULD THERE BE OTHER CULTURALLY DISTINCTIVE SCIENCES THAT WORK?

Do any other knowledge traditions deserve to be called sciences? The conventional view is that only modern sciences are entitled to this designation, so one has to examine what the criteria are that award this designation to exactly one of the many traditions of knowledge seeking about the world around us. In such accounts, science is treated as a cultural emergent in early modern Europe. Although a transformation of social conditions may have made it possible in the first place, what emerged was a form of knowledge seeking that is fundamentally self-generating; its "internal logic" is responsible for its great successes. This "logic of scientific research" has been characterized in various ways—as induction, crucial experiments, the hypothetico-deductive method, a cycle of normal science, then paradigm revolution, then normal science again. Whatever the logic attributed to scientific research, it is conceptualized as inside science, not outside science such that it is in society. Though Chinese or African astronomers may have made discoveries before Europeans, this is not sufficient to indicate that the former were really doing what is reasonably regarded as real science.[10]

For one thing, Westerners note that Chinese or African astronomy is done within culturally local projects of a sort devalued by scientific rationality, such as (in some cases) astrology, religion, or culturally local meanings of the heavens or other natural phenomena. So, whatever their accuracy, such astronomical discoveries could not be admitted as real science without permitting the possibility of assigning such a status also to the elements of astrology or Confucian religious beliefs that permeated such inquiries. Alternatively, one could say that only those discoveries of other cultures that are duplicated by Western sciences count as scientific; this has the paradoxical consequence that as Western sciences develop, incorporating insights, methods, or ontological assumptions of other cultures (intentionally or not), other cultures also (retroactively!) get more scientific. It is clear that Western cultures have had a problem figuring out how to characterize other culture's systems of knowledge about the natural world in ways that leave their favored histories and philosophies of science unchallenged. Whereas science is said to need a supportive social climate in order to flourish, the particular form of that climate is claimed to leave no distinctive cultural fingerprints on science's results of research.

Is this a reasonable position? Is the content of the successes of modern sciences due entirely to the sciences' "internal" features? Several kinds of skeptical responses to such a claim have appeared in recent years. For one, not all of the successes attributed to Western sciences are unique to it. In many cases, "what has been ascribed to the European tradition has been shown on closer examination to have been done elsewhere by others earlier. (Thus Harvey was not the first to discover the circulation of blood, but an Arabic scientist was; Paracelsus did not introduce the fourth element 'salt' and start the march towards modern chemistry, but a twelfth-century alchemist from Kerala did so teaching in Saudi Arabia.)"[11] Many other cultures made sophisticated astronomical observations repeated only centuries later in Europe. For example, many of the observations that Galileo's telescope made possible were known to the Dogon peoples of West Africa more than fifteen hundred years earlier. Either they had invented some sort of telescope or they had extraordinary eyesight.[12] Many mathematical achievements of Indians and other Asian peoples were adopted or invented in Europe only much later.[13] Indeed, it is as revealing to examine the ideas that European sciences *did not* borrow from the knowledge traditions they encountered as it is to examine what they did borrow. Among the notions "unborrowed" are the ability to deal with very large numbers (such as 10^{53}), the zero as a separate number with its own arithmetical logic, and irrational and negative numbers.[14] Historian of Chinese science

Joseph Needham points out that "between the first century BC and fifteenth century AD Chinese civilization was much more efficient than the Occidental in applying human natural knowledge to practical human needs. . . . [I]n many ways this was much more congruent with modern science than was the world outlook of Christendom."[15] Thus, other knowledge traditions "worked" at projects Western sciences could accomplish only much later. If the achievements of modern science should be attributed to its "internal logic," then evidently this logic is not unique to it.

This brings us to a second point. Nobody has discovered an Eleventh Commandment handed down from the heavens specifying what may and may not be counted as a science. Obviously, the "demarcation project" of drawing a line between science and nonscience is undertaken because it emphasizes a contrast thought to be important. Belief in the reality of this demarcation, as in the reality of the science versus pseudoscience duality, is necessary to preserve the mystique of the uniqueness and purity of the West's knowledge seeking. The self-image of the West depends on contrasts not only between the rational and irrational but also between civilization and the savage or primitive (or feminine or womanly), the advanced or progressive and the backward, dynamic and static societies, developed and undeveloped, the historical and the natural, and the rational and the irrational, among others. Thus, the sciences and their philosophies of uniquely Western rationality have been (largely unacknowledged) partners with anthropological studies of the irrational and primitive. This partnership has maintained a whole series of Eurocentric contrasts, whether or not individual scientists, philosophers, or anthropologists so intended. Through such contrasts the European Self has constructed its Other, and thereby justified its exploitative treatment of peoples it was already or about to be engaged in subjugating.[16] My point here is that even though there clearly are obvious and large differences between modern sciences and the traditions of seeking systematic knowledge of the natural world to be found in other cultures, it is useful to think of them all as sciences in order to gain a more objective understanding of the causes of Western successes, the achievements of other sciences, and possible directions for future local and global sciences.[17]

One cannot avoid noticing, moreover, that Northern science studies scholars disagree on just what are the distinctive features responsible for the success of European sciences. It is instructive to look at five accounts of Western scientific uniqueness made by distinguished and otherwise progressive Northern analysts—ones whose work has in important ways challenged conventional Eurocentric assumptions. Anthropologist Robin Horton, who has shown how African traditional thought is surprisingly similar to Western sci-entific thought, attributes the residual crucial differences to the fact that mod-ern scientific thought takes a critical stance toward tradition and is aided in this project by its rejection of magical relations between language and the world; it holds that we can manipulate language without thereby changing the world. However, as philosopher J. E. Wiredu points out, Horton under-values the extent of noncritical and dogmatic assumptions in modern West-ern scientific thought. After all, classical British empiricism has for some time been traditional thought for Western scientific communities and those who value scientific rationality. The once radical claims of Locke and Hume have become uncontroversial assumptions for us. An anthropologist from another culture might refer to them as our folk beliefs. So how accurate is it to claim that a critical approach to (all of our) traditions is responsible for the suc-cesses of modern sciences? Moreover, if science is modern in its rejection of magical relations between language and the world, scientists surely are not, Wiredu continues, since many also hold religious beliefs that invest in just such magical relations. Many commentators have noted that sacred—dare one say magical—faith in the accuracy and progressiveness of modern sci-ence is characteristic of many scientists and of the educated classes in the West and, in many cases, elsewhere, a point to which I shortly return.[18]

Historian Thomas Kuhn would agree with Wiredu's assessment that West-ern sciences are in significant respects uncritical of conventional assumptions; indeed, he argues that they are dogmatic in rejecting a thoroughgoing criti-cal attitude. However, he has explained that this scientific dogmatism is not an obstacle to scientific progress but, instead, a crucial element in its success. A field becomes a science only when it no longer questions a founding set of assumptions within which it can then get on with the business of designing research projects to resolve the puzzles that such assumptions have brought into focus. He attributes the unique successes of modern sciences to the dis-tinctive organization of Western scientific communities: "Only the civiliza-tions that descended from Hellenic Greece have possessed more than the most rudimentary science. The bulk of scientific knowledge is a product of Europe in the last four centuries. No other place and time has supported the very spe-cial communities from which scientific productivity comes."[19] One of the controversial features of Kuhn's account was his claim that the social com-munity of scientists is internal to the logic of science. The very special scien-tific communities are ones trained to follow modern science's internal logic of paradigm creation, puzzle solving with anomaly tolerance, paradigm breakdown, and then, eventually, another paradigm shift. Moreover, they create a sense of mutual trust sufficient to encourage expression of the most rigorous critical perspectives, by anyone and everyone in the community, on each other's thinking and practices. Kuhn's argument here opened the door to including the sociology of the production of what societies regarded as the very best scientific knowledge inside the "logic of science." However, one can also see that Kuhn's problematic here, his concern to identify a different, distinctive cause of modern science's successes, is inseparable in his thought from the widespread Eurocentric assumptions he articulates about the origins and virtues of European civilization.

Historian Joseph Needham, who does refer to Chinese knowledge traditions as sciences when comparing them to those of the modern West, and who would contest Kuhn's characterization of non-European sciences as primitive and the West's as uniquely descended from the Greek, proposes yet another kind of cause of the success of modern European sciences.

> When we say that modern science developed only in Western Europe at the time of Galileo in the late Renaissance, we mean surely that there and then alone there developed the fundamental bases of the structure of the natural sciences as we have them today, namely the application of mathematical hypotheses to Nature, the full understanding and use of the experimental method, the distinction between primary and secondary qualities, the geometrisation of space, and the acceptance of the mechanical model of reality. Hypotheses of primitive or medieval type distinguish themselves quite clearly from those of modern type.[20]

For Needham, what is unique to the real science of the modern West is not the attitudes of scientists on which Horton focuses, or the organization of scientific communities that appears so important to Kuhn, but a specific set of assumptions about the nature of reality and appropriate methods of research.

Sociologist Edgar Zilsel, asking why modern science emerged only in Renaissance Europe rather than in China or some other "high culture," claims that the emergence of a new social class that was permitted to combine a trained intellect with willingness to do manual labor allowed the invention of the experimental method. In slave societies or rigid aristocracies, manual labor is assigned to the lowest classes, which are also denied education. The distaste for manual labor is consequently shunned by the educated classes. Thus, no individuals are trained into both head and hand labor. Only during the emergence of a new social formation in early modern Europe, where there was an absence of slavery and of such a comprehensive and rigid aristocracy, was there a progressive culture, he implies, that gave individuals reasons and opportunities to obtain both intellectual and manual training. The new class of skilled craft workers—painters, gold and silver metal workers, lens grinders, and astronomers—obtained the very best educations in craft work available, which required new intellectual and manual abilities.[21]

Finally, Edwin Hutchins, in the context of a comparative study of "distributed knowledge" among Polynesian navigators, on the one hand, and U.S. Naval Academy navigators, on the other hand, proposes that modern science is distinguished by its ability to "black-box" information into scientific technologies and contexts. This makes it possible for people who use this knowledge to avoid having to either memorize or personally conduct the calculations they need in order to use and add to such knowledge. The U.S. navigators "distribute" the information they need for their navigational feats across the towers, cliffs, and islands they use to help locate their position at any given moment, the distinctive tasks of different sailors and officers assigned to navigational tasks, and the charts and instruments used to measure and record locations. No one individual at any time ever possesses all of the information and skills necessary to steer an aircraft carrier into the San Diego Naval Base. Navigation is a team project, and the towers, charts, and instruments are all team players. For Hutchins, it is the technologies of scientific research that are distinctive to modern Western sciences. Scientific research is fundamentally technological practice—a point to which I return in a later chapter.[22]

No doubt one could find additional features of the cultures and practices of modern sciences to which other scholars would attribute their successes. These different purported causes are probably not entirely independent of each other, and some readers will find more plausible one rather than another of such proposals. However, my point here is that there is no general agreement even among the most distinguished and progressive Western science theorists about the distinctive causes of modern sciences' successes and that the search for such an explanation and the kinds of accounts on which such scholars settle often remain tied to Eurocentric dualisms. Even if one or more of these criteria mark a significant difference between modern Western scientific practices and those of other cultures, we can ask if such a difference justifies the meanings and values that have been loaded onto the uniqueness of modern Western science.

A third source of skepticism about conventional claims for the unique efficacy of Western sciences arises from an often repeated argument in the postcolonial accounts. European sciences advanced because they focused on describing and explaining those aspects of nature's regularities that permitted the upper classes of Europeans to multiply and thrive, especially through the prospering of their military, imperial, and economic expansionist projects. Interestingly, evidence for this claim began to become more widely available through many of the museum exhibits and scholarly publications associated with the 1992 quincentennial of the "Columbian encounter" in the Americas. These materials drew attention, intentionally or not, to the numerous ways European expansion advanced European sciences.[23] A detailed account of how British colonialism in India advanced European sciences is provided by historian R. K. Kochhar. The British needed better navigation, so they built observatories, funded astronomers, and kept systematic records of their voyages. The first European sciences to be established in India were, not surprisingly, geography and botany.[24] Nor is the intimate relation between scientific advancement in the West and expansionist efforts only a matter of the distant past (or only of expansion into foreign lands, as noted earlier). By the end of World War II, the development of U.S. physics had been virtually entirely handed over to the direction of U.S. militarism and nationalism, as historian Paul Forman has shown in detail. And critics of the West's Third World development policies persistently argue that these policies have supported the further imperial practices of the West in the Third World, advancing Western knowledge and greed at the expense of the poorest citizens in the Third World—precisely those who were supposedly the most important beneficiaries of development policies. Development has brought dedevelopment and maldevelopment to the "have-nots" and economic benefits to the investing classes in the Global North.[25]

Thus, European expansionism has changed the "topography" of global scientific knowledge, causing the advancement of European sciences and the decline or underdevelopment of scientific traditions of other cultures: "The topography of the world of knowledge before the last few centuries could be delineated as several hills of knowledge roughly corresponding to the regional civilizations of, say, West Asia, South Asia, East Asia and Europe. The last few centuries have seen the levelling of the other hills and from their debris the erection of a single one with its base in Europe."[26]

These arguments begin to challenge the belief that the causes of modern sciences' achievements are to be located entirely in their inherently transcultural character. It turns out that what makes them "work," and to appear uniquely to do so, is at least partly a consequence of their focus on kinds of projects that European expansion could both advance and benefit from while simultaneously clearing the field of potential rival scientific traditions. To make such claims is not to deny that Western sciences can claim many great and historically unique scientific achievements. Instead, it is to argue, contrary to conventional views, that the very best-supported scientific claims and practices, no less than false beliefs, are caused by social relations as well as by nature's regularities and the exercise of human reason.[27]

Finally, the "internal logic" arguments are undermined by studies of how even the cognitive, technical cores of Northern sciences are permeated by historically and culturally local values, interests, discourses, and choices about how to organize the production of scientific knowledge, as we shall see in the next section.

IS MODERN SCIENCE CULTURALLY "WESTERN"?

The same accounts that have been describing the histories of other scientific traditions also show the distinctive cultural features of modern sciences.[28] These features are, for better and worse, precisely the features that are responsible for their successes, as the discussions above began to reveal. That is, the distinctive social and political history of the development of modern sciences is not external to the content of these sciences; it appears in the representation of nature's regularities and underlying causal tendencies that they produce, including in their "laws of nature" and their historically distinctive practices or methods that are at their cognitive or technical core. Particular moments in the history of science have been shaped by the intellectual, social, economic, and political concerns of their eras. Of course, this is the argument made also in the post-Kuhnian science and technology studies research. However, four distinctively Western features rarely noted in the Northern writings are persistently identified in the postcolonial literature.

First, as indicated above, which aspects of nature modern sciences describe and explain, and how they are described and explained, have been selected in part by the conscious purposes and unconscious interests of European expansion. Of course, these are not the only purposes and interests shaping these sciences—androcentric, religious, local gender, and bourgeois interests have also had powerful effects (and have also all been part of European expansion projects), as many recent accounts have shown—but they are significant. The problems that have gotten to count as scientific are those for which expansionist Europe needed solutions. Those aspects of nature about which the beneficiaries of expansionism have not needed or wanted to know have remained uncharted. Thus, culturally distinctive patterns of both systematic knowledge and systematic ignorance in modern sciences' pictures of nature's regularities and their underlying causal tendencies can be detected from the perspective of cultures with different preoccupations.

For example, modern sciences answered questions about how to improve

European land and sea travel; mine ores; identify the economically useful minerals, plants, and animals of other parts of the world; manufacture and farm for the benefit of Europeans living in Europe, the Americas, Africa, and India; improve their health and occasionally that of the workers who produced profits for them; protect settlers in the colonies from settlers of other nationalities; gain access to the labor of the indigenous residents; and do all this to benefit only local European citizens—the Spanish versus the Portuguese, French, or British. These sciences have not been concerned to explain how the consequences of interventions in nature for the benefit of Europeans of the advantaged gender, classes, and ethnicities would change the natural resources available to the majority of the world's peoples.[29] Nor were they interested in what the economic, social, political, and ecological costs to less advantaged groups around the globe would be of the interventions in nature and social relations that sciences' experimental methods made possible. Sciences with other purposes—explaining how to shift from unrenewable to renewable natural resources, to maintain a healthy but less environmentally destructive standard of living in the overdeveloped societies, to clean up toxic wastes, to benefit women in every culture, and so on—could generate other, perhaps sometimes conflicting, descriptions and explanations of nature's regularities and underlying causal tendencies.

One especially valuable regional resource available only in Europe was created through the intermingling and integration of non-European elements with each other and with resources already available in Europe in order to make more useful elements for modern science. That is, those non-European elements indicated above were not only borrowed but also frequently transformed through processes possible only for a culture at the center of global exchanges. Thus, theoretically, the map and route of European expansion could be traced in the expansion of the content of European sciences. Prior to European expansion African, Asian, and indigenous American cultures had long traded scientific and technological ideas among themselves as they exchanged other products, but this possibility was reduced or eliminated for them and transferred to Europe during the "voyages of discovery."[30]

Second, early modern sciences' conception of nature was distinctively Western or, at least, alien to many other cultures. For the resident of medieval Europe, nature was enchanted; the "disenchantment of nature" was a crucial element in the shift from the medieval to the modern mentality, from feudalism to capitalism, from Ptolemaic to Galilean astronomy, and from Aristotelian to Newtonian physics.[31] Modern science related to a worldly power in nature, not to power that lay outside the material universe. To gain power over nature would, for modern man, violate no moral or religious principles.

This aspect of non-Western knowledge systems has remained a serious obstacle to the willingness of Westerners to count such systems as "real science" regardless of how well they work to predict the regularities of the world around us. This is one reason it is so important to identify the distinctively culturally local features of modern Western sciences. If our very best sciences are also embedded in culturally local beliefs and assumptions, then the fact that other cultures' knowledge systems are similarly constituted within such assumptions cannot serve to assign them to the category of "not real sciences." Here arise a different set of questions about the causes of modern Western sciences' undoubtable successes.

Indeed, the Western conception of laws of nature drew on both Judeo-Christian religious beliefs and the increasing familiarity in early modern Europe with centralized royal authority, with royal absolutism. Joseph Needham points out that this Western idea that the universe was a "great empire, ruled by a divine Logos," was never comprehensible at any time in the long history of Chinese science, since a common thread in the diverse Chinese traditions was that nature was self-governed, a web of relationships without a weaver, with which humans interfered at their own peril. "Universal harmony comes about not by the celestial fiat of some King of Kings, but by the spontaneous co-operation of all beings in the universe brought about by their following the internal necessities of their own natures. . . . [A]ll entities at all levels behave in accordance with their position in the greater patterns (organisms) of which they are parts." Compared to Renaissance science, the Chinese conception of nature was problematic, blocking the interest of the Chinese in discovering "precisely formulated abstract laws ordained from the beginning by a celestial lawgiver for non-human nature." Needham points out that "there was no confidence that the code of Nature's laws could be unveiled and read, because there was no assurance that a divine being, even more rational than ourselves, had ever formulated such a code capable of being read." Of course, such notions of "command and duty in the 'Laws' of Nature" have disappeared from modern science, replaced by the notion of statistical regularities that describe rather than prescribe nature's order—in a sense, a return, Needham comments, to the Taoist perspective. Yet other residues of the earlier conception remain.[32] Evelyn Fox Keller has pointed to the positive political implications of conceptualizing nature simply as ordered rather than as law governed. "Laws of nature, like laws of the state, are historically imposed from above and obeyed from below." In contrast, "the concept of order, wider than law and free from its coercive, hierarchical, and centralizing implications has the potential to expand our conception of science. Order is a category comprising patterns of organization that can be spontaneous, self-generated, *or* externally imposed."[33] My point here is only that Western conceptions of nature have been intimately linked to historically shifting Western religious and political ideals.

A third feature of Western views of nature and science appears in the way peoples of European descent both distribute and account for the consequences of their sciences, as the postcolonial accounts argue. Consider first the patterns of distribution. The benefits of these sciences are distributed disproportionately to already overadvantaged groups in Europe and elsewhere, and the costs disproportionately to everyone else. Whether one looks at sciences intended to improve the military, agriculture, manufacturing, health, or even the environment, the expanded opportunities that they make possible have been distributed predominantly to small minorities of already privileged people in the West and around the globe. The costs have been borne by the already poorest racial and ethnic minorities and women located also in the West and around the globe, that is, at the periphery of local and global economic and political networks.

The causes of this distribution are not mysterious or unforeseen. For one thing, it is not "man" whom sciences enable to make better use of nature's resources, but only those already well positioned in social hierarchies. As Khor Kok Peng puts the point, the latter already own and control both nature, in the form of land with its forests, water, plants, animals, and minerals, and the tools to extract and process such resources. These people are the ones who are in a position to decide

> what to produce, how to produce it, what resources to use up to produce, and what technology to use. . . . We thus have this spectacle, on the one hand, of the powerful development of technological capacity, so that the basic and human needs of every human being could be met if there were an appropriate arrangement of social and production systems; and, on the other hand, of more than half the world's population (and something like two-thirds the Third World's people) living in conditions where their basic and human needs are not met.[34]

Not only are the benefits and costs of modern science distributed in ways that disproportionately benefit elites in the West and elsewhere, but science's accounting practices are also designed to make this distribution invisible to those who gain the benefits. All consequences of sciences and technologies that are not planned or intended are externalized as "not science."[35] The critics argue that such an "internalization of profits and externalization of costs is the normal consequence when nature is treated as if its individual components were isolated and unrelated."[36] Thus, the very ontology of modern sciences generates these inequitable effects.

Finally, even if modern sciences bore none of the above cultural fingerprints, their value neutrality would itself mark them as culturally distinctive. Of course, this is a contradiction ("If it's value free, then it's not value free"), or at least highly paradoxical. The point is that maximizing cultural neutrality, not to mention claiming it, is itself a culturally specific value. Both the reality and the claim are at issue here. Non-Western cultures, like premodern Europe, do not value neutrality in their conceptions of and interactions with nature. Thus, a culture that does is easily identifiable. Moreover, the claim to neutrality is itself characteristic of the administrators of modern Western cultures organized by principles of scientific rationality.[37] Surprisingly, it turns out that valuing abstractness and formality expresses a distinctive cultural feature, not the absence of any culture at all. Consequently, when modern science is introduced into other societies, it is usually experienced as a rude and brutal cultural intrusion precisely because of this feature, too. Modern sciences' commitments to social neutrality devalue not only local scientific traditions but also the culturally defining values and interests that make a tradition Confucian rather than Protestant or Islamic. Claims for modern sciences' universality and objectivity are "a politics of disvaluing local concerns and knowledge and legitimating 'outside experts.'"[38]

Interesting issues emerge from the discovery of the European cultural specificity of modern sciences—issues different in part from those that have arisen from the similar discovery of cultural elements by post-Kuhnian Northern science and technology studies. For example, the conventional understanding of the universality of modern science is contested in two ways in these accounts.

First, these accounts argue that universality is established as an empirical consequence of European expansion, not as an epistemological cause of valid claims, to be located "inside science"—for example, in its method. As one author puts the point: "The epistemological claim of the 'universality of science' . . . covers what is an empirical fact, the material and intellectual construction of this 'universal science' and its 'international character.' The 'universality of science' does not appear to be the cause but the effect of a process that we cannot explain or understand merely by concentrating our attention on epistemological claims."[39]

Second, a wedge has been driven between the universality of a science and its cultural neutrality. Although the laws of nature discovered by modern sciences will have their effects on us regardless of our cultural location, they are not the only possible such universal laws of nature; there could be many universally valid but culturally distinctive sciences.

> If we were to picture physical reality as a large blackboard, and the branches and shoots of the knowledge tree as markings in white chalk on this blackboard, it becomes clear that the yet unmarked and unexplored parts occupy a considerably greater space than that covered by the chalk tracks. The socially structured knowledge tree has thus explored only certain partial aspects of physical reality, explorations that correspond to the particular historical unfoldings of the civilization within which the knowledge tree emerged.
>
> Thus entirely different knowledge systems corresponding to different historical unfoldings in different civilizational settings become possible. This raises the possibility that in different historical situations and contexts sciences very different from the European tradition could emerge. Thus an entirely new set of "universal" but socially determined natural science laws are [*sic*] possible.[40]

I would only want to challenge the apparent assumption in this passage that the world's scientific knowledge consists of a static, fixed collection of claims, only some of which have been identified as scientific. As we will see in a later chapter, it is valuable instead to think of culturally diverse scientific practices that have the perpetual possibility always of producing new insights and ways of interacting with the world around us.

Other Modern Sciences?

These accounts thus provide additional evidence for the claim that other modern sciences, ones that could solve problems troubling poor and oppressed people around the world, could be constructed within other cultures. This is the argument that I left incomplete in the last section. Significant cultural features of Western sciences have not blocked their development as fully modern, according to the postcolonial accounts. Indeed, cultural features play an important role in creating these modern successes.

Moreover, one can now ask which of the cultural purposes of modern sciences that continue today to shape its conceptual framework are still desirable. Should we want to continue to develop sciences that, intentionally or not, succeed by extinguishing or obscuring all other scientific traditions, directing limitless consumption of scarce and unrenewable resources, distributing their benefits internally and their costs externally, and so forth? Furthermore, these arguments show that if culture shapes sciences, then changes in local and global cultures can shape different sciences "here" as well as "there." The sciences we have are not inevitable. We could have created very different systems of knowledge. We can still do so.

3

With Both Eyes Open: A World of Sciences

One Planet, Many Sciences?

The indigenous knowledge systems of non-Western cultures have served those cultures well in the past. As the British anthropologist Bronislaw Malinowski noted some eighty years ago, awareness of nature's causal relations was necessary for every culture's survival.[1] Could such knowledge systems flourish again today, continually transforming themselves to engage effectively with their changing natural and social environments as they have in the past? It is hard to imagine how such projects could succeed. They would have to proceed in the face of increasing poverty in the Global South created by today's forms of Western expansion. For example, Western development policies, implemented through international agencies such as the International Monetary Fund and the World Bank as well as by national organizations and transnational corporations, have instead created the "dedevelopment" and "maldevelopment" of so many non-Western cultures, including their indigenous knowledge traditions. Yet such indigenous traditions persist in sometimes impoverished and other times robust forms around the world, including in immigrant communities in Northern countries. Many of these traditions are probably too fragile to survive the continuing onslaughts of Western expansion. Valuable achievements of human knowledge of nature are disappearing from sight on a daily basis.

Modern Western sciences have helped to disempower other knowledge traditions. Yet the days are over when knowledgeable observers can continue to believe that there is or could be exactly one "real science." Modern Western sciences certainly are "international sciences," but that is not the same as being culture neutral or the only possible effective and rational knowledge system. Indeed, the "unity of science" thesis no longer looks reasonable if one is considering only modern Western sciences, quite apart from the multitude of ontologically, methodologically, and epistemologically mutually incompatible knowledge systems still to be found in other cultures.[2]

Yet the appeal of the unity ideal is difficult to resist when it is advanced by such well-known and powerful contemporary scientists as, for example, the influential author of *Consilience: The Unity of Knowledge*, Edward O. Wilson. He believes "in the unity of the sciences—a conviction, far deeper than a mere working proposition, that the world is orderly and can be explained by a small number of natural laws."[3] Moreover, some science intellectuals from other parts of the world have become alarmed at what they see as the increasing popularity of the science studies critiques of familiar philosophies of science, and especially of arguments supporting indigenous knowledge systems.[4] Most people in the West as well as many educated people elsewhere cannot imagine how alternatives to modern Western sciences could ever be effective or gain widespread legitimacy. The unity of science appears as the only reasonable response to what is claimed to be the natural condition of the world. How could any other organization of scientific knowledge and practice possibly be intelligible?

In light of this situation, it would be useful to begin to envision alternatives to the unity-of-science thesis. Of course, we can look at the historical records and see that many other traditions existed before the emergence of modern Western sciences. And, to be sure, remnants and residues of these traditions have existed even into the present day. So the project here is not simply to imagine ourselves in another knowledge tradition that is conceptualized as isolated from the rest of the world and, especially, from modern Western sciences. This would be a kind of conventional anthropological or historical project. Instead, the project here will be to envision keeping "both eyes open"—one on contemporary Western sciences and their philosophies and the other on other cultures' scientific practices and legacies. What kinds of future relations can we imagine between modern Western sciences and these other science traditions? What possibilities appear if we open both eyes, not just the one that has been fixated on modern Western sciences and technologies?[5] Such an exercise can help us to think more creatively about the possibilities and the value for everyone of living in a world of multiple knowledge systems.

We live on one planet. The scientific choices made by each culture have effects on other cultures and their knowledge systems as one culture's choices bring about changes in the environment, in animal and human bodies, and in the economies, educational systems, media concerns, and other features of societies. Of course, these days it is the choices made by the Global North that have the most powerful effects on other cultures' conditions and opportunities. Moreover, class, gender, ethnicity, religion, and history produce different and conflicting approaches to science and technology issues in the metropolitan centers, as they do in the cultures at the global peripheries. The vast majority of the world's peoples who were already economically and politically most vulnerable have had to bear most of the costs and have received fewest of the benefits of the advance of modern Western sciences and their applications and technologies.

To be sure, it would be a mistake to hold that all of the difficulties Third World cultures face at this moment in history are the doing of the Global North or its sciences. These cultures, too, have indigenous legacies of inequality and exploitation. They, too, have followed science and technology policies that turned out to be unwise. They, too, have suffered from natural and social processes that they could not figure out how to escape. The point here is that the balance sheet for both modern sciences and other knowledge traditions looks different from the perspective of the lives of the world's citizens who are economically and politically most vulnerable than it does from the lives of advantaged groups in the Global North and elsewhere. Moreover, there are good reasons to think that in some respects the perspectives of the elites are less objective.

Suggestions in the postcolonial science and technology writings can help us to imagine kinds of possible future relations between modern Western sciences and other knowledge traditions. Surely, such relations actually will be more complex than the schema below suggests, and other possibilities not considered here may turn into realities. Yet considering as many possible models as we can will illuminate the history of modern Western and other cultures' sciences. It can give rise to more fruitful questions about what kinds of sciences we could and should want in the future. Let us begin by considering four kinds of projects starting in the Global South that have been envisioned by postcolonial observers, and then turn to consider some that can start from modern Western sciences.

Projects Starting in the Global South

Of course, peoples around the world share certain needs and desires—for example, adequate food, clean water, shelter, good health, community, and protection from dangers, natural and social—though one must take care not to make a false universal claim in specifying such needs and desires, as Northerners all too often have done. Yet cultures' locations in local and global social relations make different projects important and give each access to different resources. Different cultures occupy different locations in nature's heterogeneous order; they live on prairies or islands, in cold or tropical climates, with monsoons, droughts, hurricanes, earthquakes, avalanches, sandstorms, or tornadoes. They have different interests in nature even in "the same" location. On the borders of the Atlantic some will be interested in fishing, others in using the ocean as a coastal trading route or a refuse dump, still others in mining the oil beneath its floor. They can draw on different metaphors, analogies, models, and cultural narratives to understand nature and their relation to it—conceptualizing the earth as a mechanism or, in environmental sciences, as a lifeboat or spaceship. And they tend to organize the production of knowledge about nature in much the same way that they organize the production of other social products. Moreover, a culture's location in global social, economic, and political space shapes the way it will understand each of the other distinctive relations to the production of knowledge mentioned.[6]

No one would deny that there are aspects of modern Western sciences and their cultures and practices that can and should be used to benefit all peoples living in every society. What is at issue is not that claim, but a host of others having to do with which aspects these are, how they should be used in different cultures, how the benefits and costs of their production and use are to be distributed, which achievements of other cultures' knowledge systems should also become "international," and who will make those decisions.

To begin, one must note that for the small middle classes in most Global South societies, modern sciences represent desirable resources for the ways that these groups participate in industry, agriculture, medicine, and the state organization of social life. As Third World feminists and others have pointed out, indigenous cultures and their knowledge systems often legitimate cruel and oppressive social relations.[7] Modern sciences and their philosophies provide attractive alternatives to such traditional knowledge systems. The value placed on empirical evidence and on high standards for objectivity and the impersonal, fair assessment of knowledge claims is a precious, hard-won gain when compared to many traditional ways of arriving at legitimate belief in those cultures no less than in the Global North. Moreover, participating in modern science brings the kind of higher status and increased power in important local as well as international contexts that are awarded to many things of Northern origin. Additionally, it is modern scientific practices that are demanded as a condition of economic aid by such international organizations as the World Bank and the International Monetary Fund. Thus, many citizens of the Global South whose voices reach the Global North are no more critical of modern sciences than are many Westerners. The attempt to nourish the ideal of a world of well-functioning sciences would seem to face serious obstacles in the Global South no less than in the Global North.

However, great social changes have often been stimulated by the farsighted projects of a few visionaries. Think, for example, of the good effects of the work of such people recently regarded as "kooks" as environmental whistle-blower Rachel Carson, the activists who have been slowly forcing the U.S. Food and Drug Administration to require more accurate labeling of our foods and drugs, and the gay health activists who brought about changes in the standards that the National Institutes of Health invoked for research on AIDS. Many features of the postcolonial critical analyses already express perspectives that appear in virtually every Global South culture and are rapidly gathering support in the North. Certainly, many Northerners are beginning to recognize that they benefit from more extensive and respectful discussion of the issues raised in and by these writings.

What could be appropriate relations between other cultures' knowledge systems and modern Northern sciences if we give up the unity of science goal? We do not have to think these all up for ourselves, for we can start with suggestions that have appeared in the postcolonial science studies literature.

INTEGRATE OTHER SCIENCES INTO WESTERN SCIENCES

One proposal is to integrate endangered indigenous knowledge systems into the sciences originating (so Northerners think) in the Global North. The continued expansion of Northern economic, political, social, and cultural relations is rapidly causing the extinction of many indigenous cultures, from the rain forests of the Amazon to the ice-bound lands where the Inuit traditionally hunted and fished to the urban centers where many American Indians now live. These processes threaten to lose for humanity the unique and valuable kinds of knowledge that such cultures have achieved.

> Just as forest peoples possess much knowledge of plants and animals that is valid and useful, regional civilizations possess stores of elaborate knowledge on a wide variety of topics. These stores, the results of millennia of human enquiry, were lost from view because of the consequences of the European "discovery." But now it appears they will be increasingly opened up, foraged for valid uses, and what is worthy opportunistically used. The operative word should be "opportunistically," to guard against a mere romantic and reactionary return to assumed past golden ages of these civilizations.[8]

Just as modernization pressures are reducing the diversity of plant, animal, and even human genetic pools, so, too, they are reducing the diversity of cultures and the valuable human ideas developed in them. These scientific legacies are interesting and valuable to preserve for their own sake. But they can also make important contributions to modern sciences.

This approach is by no means unique to transnational corporations such as pharmaceutical companies and to the Global South activists who insist on the governance of such processes by recognition of indigenous property rights. In the past decade such arguments have appeared even in Global North popular accounts of "endangered societies." For example, *Time* ran a cover story in 1992 reporting on the "endangered knowledge" Northern societies should gather from cultures disappearing under modernization and development pressures.

Such projects raise important questions. Of course, it would be a good thing to preserve scientific legacies about to be lost. Yet if this were the only strategy for doing so, it would offer no resistance to the eventual extinction of all "freestanding" non-Northern scientific traditions. As indicated, modern Northern sciences have always "borrowed" ideas, techniques, and materials from other cultures. So this proposal would please Northern pharmaceutical companies, for example, which are already busy integrating the knowledge about plant remedies developed by Amazonian rain forest Indians into modern medicine.

Philosophically, this project would lend false support to the unity of science assumptions. It would reinforce the ideal of one authoritative knowledge system, secure in its triumphalism and exceptionalism. This project would incorporate only those non-Northern elements that *could* be incorporated without dissonance into Northern sciences. Other kinds of valuable knowledge, which Northern science cannot recognize either as knowledge or as valuable, would have to be abandoned. This project would leave us with one global science that is distinctively the product of the Northern civilization tradition and that lacks much of the knowledge other cultures have produced and need and that could become valuable to everyone. Moreover, in abandoning the rich local cultural traditions that have given birth to and sustained other cultures' knowledge projects, such a process would lose resources for future scientific discoveries and inventions, and for human flourishing.

Finally, this criticism illuminates who have been the beneficiaries and who have had to bear the costs of "modernization" and "development" projects that supposedly were to benefit the world's already least-advantaged peoples. How should Northerners feel about extracting for the benefit of the already most advantaged societies the resources that become available from cultures that are dying as a consequence of Northern expansionist policies and practices?

DELINKING

Another proposal argues that scientific projects in the Global South should be "delinked" from Northern ones.[9] This is thought to be necessary if societies in the South are to construct fully modern sciences within their indigenous scientific traditions. Otherwise, capitalism inevitably succeeds in turning Southern cultures into markets that can increase profits for elites in the North (just as it continually extends into more and more aspects of daily life in the North). The Third World Network puts the issue this way:

> Only when science and technology evolve from the ethos and cultural milieu of Third World societies will it become meaningful for our needs and requirements, and express our true creativity and genius. Third World science and technology can only evolve through a reliance on indigenous categories, idioms and traditions in all spheres of thought and action. . . . A major plank of any such strategy should be the delinking of the Third World from the secular dynamic which institutionalizes the hegemony of the West.[10]

It makes sense at least to consider this strategy when one recognizes that there are many more "universal laws of nature" that such delinked sciences could discover if they were permitted to develop out of civilizational settings different from those that have been directed by hegemonous Northern projects. Such a delinking program could make a world of different but interrelated culturally diverse sciences.

Yet it is hard to imagine how this kind of delinking could occur. We live in a continually shrinking world, where policies and practices of one country inevitably have effects on their neighbors and on distant countries. This is especially the case for policies and practices of the most powerful countries. For example, water use, fishing practices, and military projects of one country often have great effects on others. Transnational corporations can operate outside of a nation's policies, and relatively unfettered by weak international laws. Moreover, so many contemporary problems are global in nature, refusing to recognize national borders. AIDS, SARS, and the Asian bird flu spread quickly around the globe. Atomic radiation and the effects of pesticides, acid rain, and other environmental pollutants quickly spread. Legal and illegal migrations, forced and voluntary, transport one country's problems and resources to another. Global media and Internet linkages do not respect border controls. How delinked could any culture become in this kind of densely linked world?

Of course, some cultures—in Africa and elsewhere—are already so very impoverished and powerless that they are already delinked in important respects from Northern scientific and technological traditions. If they do not offer the North either natural or human resources or markets, they have become de facto delinked in important ways from the global political economy. Yet they must still suffer the effects of such economies and their technologies on their environments, and thus on their health, life, and well-being. For example, impoverished African, Asian, and Latin American cultures are not delinked from the effects of global pandemics, environmental destruction, or their neighbors' well-armed expansionist agendas.

However, even if anything that could reasonably be recognized as social, political, and economic delinking proves impossible, attempting to delink as much as possible—even just daring to think about it—enables more critical thinking in the North and more creative strategizing elsewhere. For example, we in the North can begin to try to imagine what scientific culture and practices in the North would or should be if the South no longer provided so much of the raw materials, "laboratories," workers, or markets, voluntary or involuntary, for modern sciences and the kinds of so-called development they have advanced. What if we could not export to the least-developed cultures the North's environmentally polluting industries, toxic wastes, and sweatshop labor? What would a sustainable North look like in this respect?

There are alternatives to the world in which we live. Considering "delinking" enables more critical thought about just what can be practical and desirable forms of resistance to continued European and U.S. expansion.

INTEGRATE NORTHERN SCIENCES INTO OTHER SCIENCES

A third proposal is to integrate in the other direction, where other cultures so desire. Thus, indigenous scientific traditions around the world would be strengthened through adopting those parts of Northern sciences that they desired to integrate into their own indigenous traditions, leaving intact the distinctive local identities of such traditions.

This is not a new idea or a new practice. This kind of process has already occurred continuously during encounters between indigenous peoples and peoples of European descent. Other cultures selectively borrowed scientific and technological resources from the Europeans they encountered, as the Europeans were borrowing from them. For example, one report describes how Western medicine was integrated into local knowledge systems in China. The Japanese selectively borrowed from the European sciences and technologies that they encountered, always retaining the distinctive national identity of their project while modernizing it with the assistance of European resources.[11] Indian science theorist Ashis Nandy argues for a comprehensive program of this sort. He points out that India

> is truly bicultural. It has had six hundred years of exposure to the west and at least two hundred years of experience in incorporating and internalizing not merely the west but specifically western systems of knowledge. It need not necessarily exercise the option that it has of defensively rejecting modern science *in toto* and falling back upon the purity of its traditional systems of knowledge. It can, instead, choose the option of creatively assessing the modern system of knowledge, and then integrating important segments of it within the frame of its traditional visions of knowledge. In other words, the Indic civilization today, because it straddles two cultures, has the capacity to reverse the usual one-way procedure of enriching modern science by integrating within it significant elements from all other sciences—premodern, nonmodern and postmodern—as a further proof of the universality and syncretism of modern science.[12]

In this scenario, there would be many culturally distinctive scientific traditions that shared some common elements with modern Western sciences and, no doubt, each other. Two forms of multiculturalism would be advanced: culturally diverse sciences around the globe and diverse cultural origins (recognized as such) within each local science.

Is this proposal possible? To some extent it is already a fact, as the discussion above indicates. Moreover, it is an increasingly prevalent one as other cultures reassert the desirability and legitimacy of their own traditions in the context of decolonization projects. Additionally, there is increasing Northern appreciation and respect for multiculturalism in general, and for an end to Northern imperialism and colonialism. These days there are regularly occurring conferences around the globe focused on indigenous knowledge traditions and their legal rights. There are journals devoted to ethnobotany, ethnomathematics, and similar topics and an online journal, the *Indigenous Knowledge and Development Monitor*.[13] Moreover, at least some Northern institutions have become more hospitable to recognizing other scientific and technological traditions both at home and around the globe. For example, the University of California at Los Angeles hospital now features a Center for East-West Medicine, directed by a physician with medical degrees from both UCLA and a Hong Kong medical school. This center champions "a new health model that blends the best of Chinese medicine with modern biomedicine."[14] In cities with large immigrant populations from other cultures such arrangements have become increasingly welcome. An exploration of other cultures' knowledge traditions occurred in a traveling exhibit originating in the mid-1990s at the Ontario (Canada) Museum of Science and Technology in Toronto. Titled *A Question of Truth*, it presented some thirty-eight interactive exhibits for children and adults on the scientific and technological systems of other cultures.[15]

These transformations in policy and institutional practice suggest that modern Northern sciences will probably not succeed in eliminating other knowledge traditions, as in the older ideal of philosophies of science. At least some of the others seem able to thrive in close proximity to Northern sciences and technologies, borrowing from the latter when it is regarded as desirable and relying on their own traditions otherwise. Yet many of the marginalized cultures are not strong enough to resist the continued expansion of Northern-originated modernization. The global cognitive diversity that is such a potential resource for each and every culture will suffer as such cultures, their scientific and technological traditions, and the physical and social environments upon which they depend disappear. It could be a gift to Northern sciences as well as to those of other cultures if Northern societies even more vigorously found their own reasons to want a more democratic balance of their own and other cultures' scientific and technological projects.

SOUTHERN MODELS FOR NORTHERN SCIENCES

A fourth proposal goes even further in revaluing Southern scientific traditions. It argues that Southern sciences and their cultures can provide useful models for global sciences of the future in a number of respects. Many elements of the distinctively modern scientific ethic are unsuitable not only for economically and politically vulnerable peoples in the South and elsewhere but also for any future human or nonhuman cultures at all. For example, modern sciences' commitments to a utilitarian approach to nature, to externalizing the costs and internalizing the benefits of scientific advances, and to an ethic of increasing consumption ("development") are not ones that can support future life on earth. "Modern science has become the major source of active violence against human beings and all other living organisms in our times. Third World and other citizens have come to know that there is a fundamental irreconcilability between modern science and the stability and maintenance of all living systems, between modern science and democracy."[16] Thus, Southern scientific traditions that do not share such problematic commitments can provide models for the kinds of global sciences that our species must have in order for it and the rest of nature to survive. As two biologists put the point, Northern sciences should realistically be assessed as a transitional stage in scientific development since they destroy both the natural world and fair social relations upon which all cultures and their sciences depend.[17]

The point here is *not* that Southern cultures and their scientific traditions are all good and Northern ones all bad. Southern cultures, no less than Northern ones, have developed patterns of erroneous belief and systematic ignorance about aspects of nature that have left them unnecessarily vulnerable to life's vicissitudes. They, too, have supported entrenched patterns of social injustice. The point here is that all of us can learn and benefit from the distinctive achievements of Southern traditions.

Moreover, we can learn from the particular skills and insights that they have developed under imperial and colonial regimes. For example, some Southern societies have learned to negotiate effectively with a powerful North. The forms of multiculturalism that they have chosen or been forced to adopt give them valuable knowledge about how to live in a world different from that of elites in the North. They cannot afford the illusion that they are dependent on no other culture. They cannot get away with imagining that they can take what they wish from nature and other peoples. Nor can they expect respect beyond their borders for the idea that they are the one model of the uniquely and admirably human or that their ideas are uniquely and universally valid. Southern scientific traditions can offer valuable models for global sciences here, too.

In this scenario, presumably, Northern groups would integrate into their sciences and culture precisely those Southern cultural elements that would transform modern sciences into more effective ones for both better sciences and social justice tendencies in the North to flourish. In contrast to the first proposal, it would be precisely some of the elements of Southern cultures most incompatible with modern sciences and their philosophies that would be valued: the Southern forms of democratic, pacific, life-maintaining, and communal tendencies, where they exist, that are so at odds with imperialistic, violent, consuming, and possessively individualistic ones that critics find in Northern sciences and culture. To be sure, the former are not always well practiced in Southern cultures. Nor are they absent from Northern cultures, as the postcolonial critics are perfectly aware and always caution. In this fourth scenario, there would be many culturally different sciences, each with culturally diverse origins. But central among the elements most valued in each case would be those that advance cooperation, democracy, the richness of indigenous achievements, and sustainable development.

Is this a real possibility? Will people of European descent be able to accept the idea that their democratic traditions, in which scientific procedures play a central role, are not the only viable ones?

It is time to turn to the other half of this chapter's thought experiment in order to identify what those living in the North can contribute to the development of sciences that have greater validity and are less imperial.

Projects Starting in the North

There are unique contributions to viable future sciences that can be made by those of us who value many features of the European tradition but are opposed to the history of modern sciences' service to militarism, profiteering, and social injustice. Progressives in the North do not have to retreat to stoicism when the topic of science is raised. We do not have to regard our only appropriate responses to criticisms of Eurocentrism as defensive ones. Instead, we can recognize the opportunities and challenges in critically retrieving and "modernizing" the best in European and U.S. cultural traditions for sciences that are suitable participants in an emerging postcolonial world in which respect for the achievements of many cultures is a legitimate ethic.

One important project is to add to our local environments—our classrooms, faculties, conferences, syllabi, footnotes, policy circles, television interviewees and interviewers, newspaper reports, and the like—the voices and presences of peoples whose cultures have borne more of the costs and received fewer of the benefits of Northern sciences. Note that simply adding "diverse" peoples to scientific sites has evidently never in itself been sufficient to exert the kind of transformation called for in the North by postcolonial science studies. "Add diversity and stir," to paraphrase a feminist saying about such practices, will be insufficient to eliminate Eurocentric sciences and their philosophies. Here are several contributions that appear fairly obvious though not uncontroversial.

We can relocate the projects of sciences and science studies that originate in the Global North and those originating in other cultures on the more accurate historical maps created by the new indigenous knowledge and postcolonial studies. We can abandon the familiar map charted by Eurocentric accounts of the mainly European and U.S. history of science, and by anthropological accounts of other cultures' superstitions and false beliefs. This project will require rethinking what it is that sciences and science studies should be describing and explaining, and how they should do so. In what ways have the existing projects in physics, chemistry, engineering, biology, geology, medicine, and environmental and other sciences been excessively contained by Eurocentric assumptions and goals? How have the conceptual frameworks and practices of Eurocentric philosophies of these sciences guided and made them appear not only reasonable but also the only such reasonable kinds of sciences?

Other equally important transformations are necessary to understand such a map—indeed, to create it in the first place. We are not used to keeping a clear focus on both Northern sciences and those of other cultures. The natural sciences have lacked the kinds of critical historical, philosophical, sociological, and ethnographic resources necessary for them to identify Eurocentric beliefs that limit their understanding of the natural world and the history of modern Northern sciences and other cultures' knowledge systems. Simply reporting other cultures' achievements will not provide these resources. As many contemporary observers of Northern sciences and their philosophies put the point, these Northern sciences and their philosophies are epistemologically underdeveloped.[18]

Scientists and humanists have usually spoken as if intellectual life should be divided between their two kinds of projects, as if the sciences and humanities are parallel or comparable kinds of human projects.[19] But they are not—at least not in the contemporary dominant form of scientific education and practice. Such an imagined universe of intellectual resources lacks the critical social sciences and, in particular, critical social studies of science and technology. These latter are a kind of science of science, mapping the relations between the social and intellectual features of scientific institutions and their practices and cultures. We need "science criticism" in the way we need the fields of literary or art criticism. The post-Kuhnian social studies of science and technology provide many of the resources for such a field, though these accounts often lack the kind of overt political analysis one can readily find in literature and art criticism.

Someone well educated in the humanities is expected to have a good training in literary, artistic, dramatic, and other forms of humanist *criticism*—in the "history, theory, and sociology" *about* these arts—but we do not expect them to be accomplished poets, sculptors, or playwrights. And this humanist critical education is not considered to be a lesser field than the creation and performance of literature and the arts. It is not an introductory project of explaining "arts for nonmajors"; it is an equal and different project with its own principles and goals—one through which poets, sculptors, or playwrights can themselves gain greater resources through exposure to the achievements and limitations of past efforts in their fields but one that sets its own goals and projects, apart from those of poets, sculptors, or playwrights.

It is a kind of comparable program for the sciences needed by scientists, educated citizens, and old-fashioned historians, sociologists, and philosophers of science. Such science studies, I propose, should be introduced into all science education programs both inside and outside classroom contexts. Existing science programs are supposed to instill in students a commitment to the most rigorous criticism of traditional assumptions, but the postcolonial accounts show that Eurocentric assumptions have blocked a crucial range of such criticisms. The sciences have not been scientific *enough* to chart the complex relations between their supposedly purely natural objects of study and the economic, political, social, and cultural assumptions and priorities that they and their cultures bring to scientific projects.

There are many purportedly reasonable answers to the question of why no such field of science studies exists already within science departments: "The history and sociology of physics is not physics," "Students don't need the history of physics to do good physics," "The methods of history and sociology are not really scientific," "Not enough time in the curriculum," "Where are the faculty to teach such courses?" and so on. And, of course, some may reasonably object to my drawing this parallel between the arts and sciences at all since artists, in contrast to scientists, are not primarily trained in universities—or, at least, were not until recently. My point here is that although we can understand and sympathize with these answers, they are *no longer reasonable enough*. Failing to locate any significant critical studies of the sciences in universities, in general education requirements, and, especially, in science departments indicates to students that no one thinks such studies important for learning to do science or for making reasoned decisions about scientific issues in public life. This is unfortunate, since, as the postcolonial accounts show, philosophical, sociological, and historical assumptions form part of scientific understanding *about nature*. Scientists unknowingly use distorting cultural assumptions as part of the *evidence* for their results of research when they are directed to ignore the social context of the "nature" they study and of the history and present practice of scientific inquiry. They, and we, need to learn how to identify cultural features in our scientific assumptions and how to sort out those that encourage unaccountability, irresponsibility, and limitations on knowledge from those that do not.

Of course, the kinds of philosophy and social studies of science needed for this project are not widely practiced. Unfortunately, even many practitioners of the new post-Kuhnian social studies of science and technology lack the empirical knowledge, social theory, and critical perspectives required to come to terms with postcolonial histories and critiques. Yet the numbers of those with such training are increasing. National and international initiatives could speed up this process of reeducation.

Imagine if every science department contained the proportion of "science critics" to scientists that there are of literary critics to creative writers in English departments. Imagine having scientists, science policy makers, and the rest of us educated in "The Role of Biology, Chemistry, and Physics in the Expansion of the Modern European Empire—and Vice Versa"; "Chinese (Islamic, South Asian, African, Indigenous American, and the like) Sciences: Past, Present, and Future"; "The Gender Effects of the Scientific Revolution in European Expansion"; "From Craft to Factory Production of Twentieth-Century Science: Benefits and Losses"; "Objectivity as Ideal and Ideology"; "The Science and Political Economy of the Human Genome Project"; "Science and Democracy: Enemies or Friends?"; and, especially, a course on the meanings and effects that our scientific projects come to have that we never intended. (Students of Jacques Derrida will have no problem producing a title for this course.)[20] Here would be a start on education that could vastly improve the empirical and theoretical adequacy of modern sciences and their philosophies, as well as their politics.

We must also disseminate such accounts both through preprofessional science education and in more general educational sites and practices. For example, these accounts should appear in the diversity-focused U.S. and global history texts currently being produced for elementary and high school students in the United States, in media accounts reaching the general public, and in journals, conferences, museums exhibits, and other forms of public education. These anti-Eurocentric accounts of a world of sciences should be part of general literacy expectations.

Finally, we can think of these kinds of tasks as a progressive project for constructing fully modern Northern sciences that creatively develop key elements of the Northern cultural legacy—that, one could say, modernize them.[21] Our Northern sciences today are not, it turns out, quite as modern as they could be insofar as they entrench traditional Eurocentric "superstitions" and false beliefs about the achievements of both Western sciences and other cultures' knowledge systems. One striking feature noted earlier in some of the Third World analyses is that they propose what we can think of as *principled* ethnosciences when they contemplate constructing fully modern sciences that conscientiously and critically use their indigenous cultural legacies, rather than— as they point out—only European ones.[22] Those of us who value some of the conceptual features of the Northern tradition can similarly strengthen notions of objectivity, rationality, and scientific method—notions central not only to our scientific tradition but also to such other Northern institutions as the law and public policy.

Paradoxically, these postcolonial analyses that can appear to come from outside modern science are also very much inside its historical processes, as I have been arguing throughout. They are exactly what is called for by its conventional goal of increasing the growth of knowledge through critical examination of cultural superstitions and unwarranted assumptions.

*   *   *

To conclude, envisioning future relations between, and projects for, both modern Northern sciences and other cultures' scientific and technological traditions can lead to far more accurate and valuable understandings not only of other cultures' scientific legacies but also of fruitful but unexplored possibilities in Northern cultures and practices. Moreover, it can provide resources for the greatly expanded accountability practices for which so many critics have called. Eurocentric philosophies of modern Northern sciences assume there is no reason to explore or reflect on any scientific traditions but their own. They leave us blind in one eye that could be critically examining other traditions. And they leave us with distorted vision in the other eye, which can see only the Global North and only from within its own self-serving mythologies.

4

Northern Feminist Science Studies: New Challenges and Opportunities

The early days of a social movement are always exhilarating as new issues and perspectives explode out of the tumult of ever changing political and intellectual experiences. Feminists had addressed issues about discrimination against women in and by the sciences since the late nineteenth century. The women's movements of the 1970s produced new energy and new directions in such work. It occurred in the context of the emerging field of social and cultural studies of science and technology. These studies took their lead from the new social histories of modern science (instead of only intellectual histories), the new sociologies of knowledge (instead of only of error or of scientists), and critical perspectives on science's involvement in militarism, ecological destruction, and the "rationalization" of labor produced in the radical political movements that swept across the United States and Europe in the 1960s. In the feminist movements, Liberal, Marxist, and Radical political philosophy traditions jostled for political and intellectual space, and the emerging voices of women of color and lesbians raised issues for everyone. In this heady context, feminist science studies projects headed off in different and sometimes conflicting directions.

Today, the field can seem to have settled into the kind of more sedate mopping-up projects characteristic of mature intellectual and social movements. Indeed, important such work is under way. But I think this is only part of the story. Here I shall try to show how feminist work has moved into deeper and more consequential examinations of the ways in which modern sciences exhibit an "integrity" with their historic moment (to borrow a phrase from Thomas Kuhn). These accounts can seem to challenge hallowed Western self-conceptions even more deeply than the early feminist science studies work. Feminist work has also provided bold alternatives that set a higher standard for mainstream science studies while it also serves feminist goals. Moreover, this work teeters on the brink of plunging into two kinds of significant shifts in focus. One is engagement with the emerging field of multicultural and postcolonial science studies. The other is a deeper and more realistic understanding of how scientific research is actually organized today in a global economy vastly different from that of even two decades ago, and of the ambiguous consequences of this shift for social justice projects. The first of these two issues will be pursued in later chapters.

Here I focus primarily on recent work in feminist studies of the natural sciences. However, as we shall see, there are good reasons to regard the nature with which the sciences engage as always already configured as social; the natural sciences and their philosophies should no longer be able to get away with regarding social analyses of scientific practice as irrelevant to understanding either nature or "real science." Indeed, those sociologies, ethnographies, political economies, and histories that constitute the social studies of the natural sciences should be reconceptualized as a crucial part of inquiry into nature's order—that is, of the natural sciences. The natural and the social cannot be kept apart from each other as conventional philosophies of science, and the natural sciences themselves, have assumed.

Feminist science studies has been immensely controversial from the very beginning. One reason is that Northern modern sciences and technologies, and especially their way of conducting research, their "logic of inquiry," are central to the ideals of modernity, democracy, and social progress. These ideals help to constitute Northern individuals' and social institutions' identities and conceptions of what are legitimate and important missions. Challenges to these ideals on behalf of women can appear impertinent, arrogant, wrongheaded, and deeply disturbing. Such challenges focus on philosophies of science, which were assumed to be immune to this kind of social criticism, and also on attempts to get women's issues into scientific and technological projects that one would assume would be more welcoming to them, such as mainstream environmental movements, medical and health research, and Third World development policy (a topic to which I return in a later chapter).

Yet the situation is more complex. At the same time, not all feminists in the world want to criticize Northern modern sciences and their philosophies. These sciences, especially as framed by positivist assumptions about a preexisting world out there available for undistorted reflection in the sciences' claims, are perceived to offer yearned-for resources to women and men in cultures still ruled by premodern forms of inequality. Positivism is not the major problem for many Third World women that it is for women in the North, as Uma Narayan argues. And Northern feminist kinds of criticisms of science end up supporting dangerous fundamentalist and nationalist politics in countries such as India, according to Meera Nanda.[1] For feminists in these cultures to focus critically on problems with a positivist philosophy of science or with Northern sciences more generally can leave them complicitous with their cultures' support of traditional forms of gender and class inequality. It can also lead them to neglect the issues that are important to women and feminists in those cultures.

As these introductory comments indicate, there is and can be no single monolithic feminism or, thus, preferred way to do feminist or gender analyses. Feminism is heterogeneous; its agent or subject of knowledge and history, its ideal knowers and historical actors, is plural and decentered in contrast to the singular, "centered" Rational Man of Enlightenment thought. This is because the experience of women in different social locations can generate different distinctive insights about dominant social institutions and their practices and cultures. Moreover, many feminist research projects are "mission directed," that is, they are designed to produce solutions to pressing economic, legal, medical and health, political, or other problems that women encounter in some particular social context. Thus, these projects do not aim to discover value-neutral, transcendental truths. Yet they do produce theoretical knowledge through such research. Indeed, for feminism, theoretical knowledge is also action in the world: how we conceptualize the world around us changes how we will interact with it and how others will interact with us. Conceptual practices are always already social and political. Thinking is also a form of doing. To put the point another way, feminist knowers, the world with which they interact, and new perspectives on that world are all simultaneously brought into existence through feminist research practices.

The term *gender* also has diverse meanings and uses, and the choices of when to use it and what to mean by it are always controversial. As a start, we can note that the term can refer to the objects of empirical study that inquirers observe, that is, to men and women or to the social structures or meaning systems that produce and reinforce gender discrimination and inequality. It can also refer to the analytic framework that researchers bring to their inquiries— to the study of how gendered individuals, social structures, and systems of meaning are socially produced and maintained. In all of these cases gender should be understood as always in a mutually constituting relationship with class, race, ethnicity, sexuality, and other structural and symbolic social systems. Thus, gendered identities, social relations, and gender analyses are all dynamic, historical phenomena, changing as the world and human understandings of it also change.[2]

Feminist Science and Technology Studies in the Global North

In Europe and the United States these analyses have been produced under the influence of more than three decades of women's movements and postpositivist social studies of science and technology. The latter have sought to show how modern sciences bear the fingerprints of the cultural, social, economic, and political values and interests of their particular time and place. They have a social integrity with their historical eras, as historian Thomas S. Kuhn put the point. Feminists have argued that this includes an integration with the gender relations of their historical eras no less than with class relations, nationalisms, militarism, and other social projects. Science studies has enabled far deeper and more innovative criticisms of the androcentrism of science, medicine, mathematics, engineering, and their philosophies than was possible for earlier generations of women who thought critically about the sciences. Yet recent feminist analyses have extended and deepened the understanding of what is social about the natural sciences; indeed, they have in this respect advanced past prevailing accounts in both mainstream science studies and in the philosophy of science.

One familiar way to organize the immense diversity of feminist approaches to science even when our horizon is restricted to Europe and North America is to identify main themes within five research directions. Feminists have focused on sexism and androcentrism in the social structure of science, in research designs and results, in science education, in the applications and technologies scientific research makes possible, and in the epistemologies and philosophies of science. This last topic requires a chapter of its own—the next one. Chapter 6 looks at these and other issues from the standpoint of women's lives in the Global South. The following sections look at the first four research directions.

SURVIVING DISCRIMINATORY SOCIAL STRUCTURES

Nineteenth-century feminists had already complained about discrimination against girls and women in the social structure of science, mathematics, medicine, and engineering. Women's caucuses in natural and social science disciplines and women's organizations in universities and in industry, some founded in the 1920s but most emerging since the 1960s, have carried on these campaigns.[3] These have been joined more recently by programs supporting the recruitment and retention of women scientists sponsored by the National Science Foundation, the American Association of University Women, other U.S. organizations, and various European Union scientific commissions.[4] It can be easy for people outside the sciences to forget how radical it remains for women scientists even to come together specifically as women scientists, to identify themselves professionally by their gender, whereas the self-image of their male colleagues remains one of individual experts whose particular biological or cultural identity is irrelevant to both the fact of their expertise and its content. After all, whatever women's male colleagues may think about the proper gender of scientists, they do not publicly identify themselves as "men scientists."

Today, when the formal barriers against women's access to science and engineering education, degrees, publication, lab appointments, and membership in scientific societies are finally illegal in Europe, the United States, and many other parts of the world, it remains challenging to identify and then eliminate powerful continuing sources of discrimination. The 1999 Massachusetts Institute of Technology's report on women science faculty at MIT created shock waves in many elite science and engineering departments as it revealed the ways in which society's gender norms, including expectations of women's obligations to family, continue to support discrimination in subtle ways against both senior and junior MIT women faculty. All of these scientists denied that they had ever experienced gender discrimination, which they identified with the practices that are now illegal. After all, these women had achieved positions at MIT. Yet the younger ones spoke of persistent difficulties negotiating between family and work obligations, and the senior scientists spoke of less access to resources—lab equipment, the best students, and so on—than their male colleagues. By now hundreds of projects in the United States have attempted to develop strategies to recruit and maintain girls and women in the natural sciences, especially through education projects.[5]

In the developing world, to skip briefly to a topic of a later chapter, the lack of economic resources and social services for families ensures that girls' domestic obligations will cause them to drop out of school long before they could gain any science education or, for many, even achieve basic literacy. And some parts of the Third World are so racked by militarism and insufficient responses to AIDS (in both of which Northern scientific and technological policy are implicated) that access to basic literacy is not high on lists of national priorities. On the other hand, many countries in eastern Europe and the Third World have far higher proportions of women on science faculties and national science policy agencies than do the United States and western European nations. This usually is not the result of feminist activism. Instead, the causes of such variation are to be found in national scientific and technological policy, differing religious and cultural patterns of gender segregation,[6] and different opportunities available to national science projects in the global political economy.[7]

Meanwhile, interest continues in the history of those women who have made important contributions or participated in significant changes in the sciences, math, and engineering. For example, recent books have reported the history of women scientists in the U.S. Navy in World War II, in post–World War II mathematics, and in the rise of science in the American West. A new biography of the infamously mistreated Rosalind Franklin appeared in 2002.[8] In a recent essay, biologist and historian of biology Ruth Hubbard returned to Franklin's case to identify how James Watson's gender politics shaped the history of science even more deeply than most observers had initially perceived.

> Watson used sexist stereotypes to obscure what should have become a scientific scandal. As Franklin's friend Anne Sayre recognized the moment she began to read *The Double Helix*, the creation of "Rosy," the humorless, dowdy, castrating female who, rather than help her dedicated male "superior" Wilkins, as she was meant to do, insists on imposing her own ideas, has the function of getting the reader not to notice that Watson and Crick had access to Franklin's unpublished data while they made their biology-shaking discovery.[9]

They gained access to these data without Franklin's permission. Watson could get away with telling the story of his research only the way he did, that is, by making fun of Franklin as a woman and boasting about the success of a project that belonged, as he saw it, to him and Crick alone, because Franklin had never been aware of the theft of her data and had long since died when Watson wrote his account. Hubbard is arguing that what can appear to be merely the mean and annoying sexist comments of a scientist, of no interest to understanding how scientific research is actually done, in this case functioned to obscure what in fact was a litigable crime: Watson and Crick stole Franklin's research results, without which they could not have been awarded a Nobel Prize. Indeed, as Hubbard points out, without that theft Watson would not have been in a position to regale readers at all with his sexist stories about Franklin.[10]

Finally, the presence of women of color in Northern feminist science studies remains minimal. For a dialogue reflecting on the costs of this absence to feminist science studies see "A Conversation on Feminist Science Studies" between Harvard historian of science, medicine, and public health Evelynn Hammonds and University of Massachusetts–Amherst biologist and women's studies scholar Banu Subramaniam.[11] Of course, there are plenty of women of color in science labs, mostly working as technicians or in other low-paid and low-status jobs. And a very few, along with their equally scarce brothers of color, are entering the top ranks of university science departments and federal and industrial laboratories. Yet the vast majority of these scientists of color are foreign born. U.S.-born African Americans, Latinos, and American Indians are about as scarce in the top ranks of science, technology, mathematics, and medicine as they are in the upper echelons of U.S. society more generally.

SEXISM AND ANDROCENTRISM IN THE RESULTS OF RESEARCH

Building on early concerns of the women's movement, feminists have continued to make important empirical and theoretical breakthroughs that correct the representations of women, men, nature, and social relations in the results of scientific research. These have often brought about significant policy changes, though in other cases even compelling empirical evidence has been unable to overcome continuing resistance from both scientists and policy makers. Both ignoring women and blaming them for the ills they experience are just too deeply ingrained local and global practices, and too costly for science and policy to abandon, for scientists and policy makers actually to engage with the empirical evidence feminist research has produced. Here are some of the most significant areas where recent feminist work has both built on yet also deeply challenged earlier feminist work as well as culturewide ways of thinking.

Our Bodies, Ourselves: The Boston Women's Health Guide was published in 1970. It has gone through four subsequent editions, sold nearly four million copies, and been translated or culturally adapted into some nineteen languages. This pioneering product of the early women's health movement revealed the ignorance that had directed physicians' interactions with patients as well as policies of the medical-industrial complex. Contrary to conventional warnings about the inevitable bad effects of politics on the growth of knowledge, it took a coalition of feminist scientists and political activists to launch this new focus of research. Obviously, some kinds of politics can advance the growth of knowledge.[12] At the same time, feminist biologists began to criticize sociobiology's claims about the naturalness of women's subordination to male domination. One group organized "genes and gender" programs at annual meetings of the American Association for the Advancement of Science and produced some of the first readers on gender and science.[13] Anne Fausto-Sterling's *Myths of Gender: Biological Theories about Women and Men* is one of the most influential analyses by a feminist biologist. This book has been widely used in medical education, as well as in the women's movement itself.

It has been more difficult to bring into feminist focus science projects concerned with racial issues, or, rather, with the intersection of race and gender issues. This is no doubt in large part because of the low representation of women of color as scientists and in feminist science studies. Fausto-Sterling was also one of the first Northern scholars to bring a comprehensive set of racial issues fully into feminist focus.[14] Her current work argues for reconfiguring the conceptual framework of biology so that gender and race can be understood in a more realistic manner. "We need . . . to develop the habit of thinking about genes as part of gene-environment systems, operating within networks that produce new physiologies in response to social conditions. In this view, bodies are not static slaves to their biology. Rather it is our biological nature to generate physiological responses to our environment and experience. We use genes to produce such responses."[15] In a new series of analyses Fausto-Sterling shows how the lack of such a gene environment–system framework consistently distorts and obscures the issues in biological, medical, and public health projects.

For some two decades Donna Haraway's powerful work has also consistently focused on the intersection of racial and gender issues in scientific research in local, national, and global contexts. She too has insisted that science can never study a nature that is conceptualized as outside culture, for nature always appears to us only as already located in cultural projects and discourses. Her examples have come primarily from primatology, but also from other kinds of animal studies. Most recently she has asked "how might an ethics and politics committed to the flourishing of significant otherness be learned from taking dog-human relationships seriously?" and how might accounts of dog-human worlds reveal the importance of history to the "naturecultures" around us?[16]

Fausto-Sterling has led the way in yet another field. Her influential 1993 essay "The Five Sexes: Why Male and Female Are Not Enough" followed by *Sexing the Body: Gender Politics and the Construction of Sexuality* have brought renewed attention to the history of and present practices in scientific studies of sexuality. These two texts have also played an influential role in supporting the emergence of an intersex (formerly known as hermaphrodite) movement, and of transgender and transsexuality movements. Thus, they have taken an important place in the field of lesbian, gay, bisexual, transgender, and Queer studies. Such fields and movements provide startling new perspectives on one of feminism's persistently important topics: gender differences and their relation to sexual differences. Consequently, the issues these fields and movements raised cannot be dismissed, as they too often are, as of little interest to feminism or to women's issues. Other recent studies of the sciences of sexuality have focused on the history of peculiar notions of the clitoris, the refusal to recognize the clitoris in female hyenas, strange notions of the female orgasm, and the invention of and medically recommended uses for the vibrator.[17]

Another fast-moving area of feminist science research focuses on environmentalism. It has arrived at a cutting edge of science studies as it shows the necessity of shifting paradigms in public health, political economy, philosophy, science, and ecology, as Joni Seager reports in a recent review essay.[18] Any account of feminist environmentalism must begin with its early subfield, ecofeminism. Ecofeminism produced fresh and exciting debates that had a powerful influence not only on the then nascent feminist and mainstream environmental movements but also on U.S. feminism more generally. Ecofeminism raised issues of the separation of spirituality from politics, of female-male and nature-culture dualisms and their links, and debates about separatism, identity, and women's activism. It turned nuclear power and strip mining into women's issues and presented a distinctive feminist approach to the peace movement. Confusingly, the term has come to signify not just this particular set of environmental concerns, but the rest of feminist environmental issues also, which in fact are different from and often in conflict with ecofeminist assumptions.

Seager proposes that although ecofeminism appears to have lost its energy, there are four lively fields of recent feminist environmental work, all of which build on earlier ecofeminist projects but also move beyond the now stale debates ecofeminism so importantly first raised. In the analyses of the oppression of animals, feminists have made important contributions to the deterioration of traditional boundaries between humans and animals, pointing out how such boundaries reprise racial, sexual, and gender hierarchies. By now there is compelling empirical evidence that animals, too, feel pain; have consciousness and social awareness; express abstract emotions; feel loss and deprivation; demonstrate cognitive skills, curiosity, and problem solving; and are motivated by "intelligence," not just by instinct. Moreover, a feminist-informed theory of animal rights challenges the misogyny and androcentrism of leading animal rights theorists and activists, such as Peter Singer and Tom

Regan. While emphasizing the commonalities between humans and animals, the feminist theories, unlike those of Singer and Regan, also value difference, refusing to value "animals only to the extent that they meet or mimic human tests of 'intelligence' and behavior." Thus, they refuse science-based identity categories. Furthermore, the feminists have exposed the gendered assumptions and perceptions that underlie human treatment of nonhuman animals. Finally, Vandana Shiva has pointed to troubles with the high value that "border crossing" has been given in much feminist and antiracist work. She argues that mad cows are also "cyborgs" and that bioengineering is another kind of border crossing.[19]

Second, feminist work has made crucial contributions to linking environmental and health issues, which were rarely linked until recently. Here this work challenges mainstream environmentalism, which long tended to ignore such issues, as well as urban and social environmental issues more generally. A well-publicized series of environmental disasters in the 1970s and '80s—the ozone hole, chemical industrial disasters at Seveso (Italy) and Bhopal, nuclear disasters at Three Mile Island and Chernobyl—linked health to environmental issues. Yet it was the women's health movement that raised the mundane, everyday issues of environmental causes of the increase in asthma and lead poisoning, the epidemic in breast cancer, endocrine disruption, and the links among ecology, health, and maldevelopment. Shiva argues that "the 'environment' is not an external, distant category. . . . The 'environment' for women . . . is the place we live in and that means everything that affects our lives."[20] When feminists have documented the toxic effects of daily environments, they have earned public recognition for these ills. Thus, they have played the "proof game" that scientific legitimacy has required. Yet environmental conditions cannot be proved; environmental research cannot escape uncertainty. An alternative strategy has been to advocate the "precautionary principle," "one of the most radical developments in global environmental thought."

> In a series of interlinked doctrines, the precautionary principle assets that public and private interests have a positive obligation to act to prevent environmental/health harm before it occurs; that the indication of harm, rather than "proof" of harm, should be the trigger for action; that the burden of proof needs to be shifted to the front of the chain of production (the presumption of safety should be tested before potentially harmful substances are released into the environment rather than waiting to test for harm after the fact); and that all activities with potential health consequences should be guided by the principle of the least toxic alternative.[21]

In two additional related areas feminist environmentalism has made distinctive contributions to global thinking and policy. It has insisted on the importance of gender issues at the convergence of theory and activism on the environment, global structural power, and the critical social studies of science. And it has persistently criticized the practice in the population debates of blaming women for environmental problems. By now, the evidence against attributing poverty and the depletion of natural resources to "overpopulation," and in particular Third World women's fertility, should be regarded as uncontestable. The evidence conclusively shows that the global political economy persists in maintaining Third World poverty and First World overconsumption of resources and production of toxicity. Yet, Seager argues, "mainstream environmental analysis—both in the popular press and imagination and in the official policy positions of many of the major U.S. and European environmental groups—detours neatly around these structural factors to place the blame for global environmental ills instead on the fertility of women in the third world."[22]

Two further issues have been important in thinking about eliminating sexist and androcentric research and its results. One we have raised already: how much effect has the now sturdy history of feminist critiques of sexism in the results of research in the natural sciences had on these sciences themselves? An at least partially distinct issue is whether women scientists have been influenced by the women's movement to pursue different questions in the sciences or to do science differently in some other way or both. Donna Haraway had examined the latter question with respect to primatology in the 1980s, and Alison Wylie had asked similar questions about women archaeologists. Evelyn Fox Keller had also addressed this question in her influential study of biologist Barbara McClintock. Londa Schiebinger has more recently focused on changes in the sciences more generally in *Has Feminism Changed Science?* and in three recent essays she brought together in a special issue of *Signs: Journal of Women in Culture and Society* focused on gender and science.[23]

The answer to each of these questions is yes and no; moreover, the details of how women and feminism have changed the sciences vary in different sciences. One striking commonality, however, is that women scientists who actively refuse to be labeled as feminists nevertheless often do ask different questions and do science differently, as Haraway, Keller, and Wylie had emphasized in the early studies. Partly, this seems to be a consequence of their often different social locations in the production of science. Excluded by "the boys" from the big teamwork research projects, they tend to develop "niche" projects that they can do with minimal assistance from peers.[24] Two new readers are designed to build bridges between feminist science studies theorists and women in science.[25]

Meanwhile, criticisms of sexist and androcentric methods and results of research in history and the social sciences had begun to appear by the early 1970s.[26] Since the natural sciences have social histories, this work has also influenced gender-focused histories, sociologies, and philosophies of science and technology. It has even shaped studies of sciences one might presume to be most immune to cultural influences, such as physics and astronomy at the origins of modern science, Boyle's chemistry, early-twentieth-century physics and biology, contemporary high-energy physics, and molecular biology.[27] Elizabeth Potter's study of Boyle is perhaps the most striking of the recent contributions to this genre. Whatever inroads feminist analyses might make to biology, health, and the "softer sciences," critics insist that Boyle's laws, like Newton's, Einstein's, and other well-defended "hard science" claims, are immune to such analyses. Yet it was Boyle's class and religious interests, as well as his well-known antipathy to women, that led him to choose a mechanistic instead of an animistic "law," where animism was associated with women, to account for his data, Potter shows. The data could equally well support either hypothesis, and Boyle never clearly refuted the animistic hypotheses offered by Franciscus Linus. Potter's account is doubly radical. First, she shows how it was local cultural assumptions and beliefs that did in fact shape the assertion of Boyle's law. Second, she does not claim that this is thereby "bad science." Instead, she makes the far more radical claim that it remains "good science." In line with the critical science studies arguments, it is the discovery of the values and interests embedded in good science that reveal the inadequacy of mainstream epistemologies and philosophies of science.

To conclude this section, feminists have continued to demonstrate how the cognitive, technical cores of modern sciences are no less a part of their eras' discriminatory social formations than are the more obvious discriminatory practices of scientists and their institutions and cultures. And they have raised provocative new issues in many sciences as well as in the social studies of science.

SCIENCE EDUCATION

Early equity approaches to remedying girls' and women's underrepresentation in the sciences and engineering assumed that these groups were deficient in the abilities and talents necessary to compete for careers in these fields. They had "math anxiety," did not like dissecting frogs, and were lacking in analytic skills. More recent work has shifted the focus to deficiencies in pedagogy, curricula, and the goals of both science and science education.[28] One illuminating approach has been the emergence of a critical focus on how girls who love science negotiate their identities in the context of the masculinized culture of science and science education. It turns out that "doing science" becomes a way of constituting certain kinds of gendered social identity. Girls and women who love doing science thus are faced with negotiating their gendered identity in a context hostile to the association of science with femininity and womanliness. Moreover, insofar as they are encouraged to develop what their mostly white and middle-class teachers regard as appropriately feminine traits, they are disabled from succeeding in fields that have associated succeeding in science with distinctively masculine traits. Yet girls have also used their love of doing science to constitute distinctive feminine identities, and, as Nancy Brickhouse has shown, in different ways for girls of different racial and ethnic identities.[29]

This work on education also offers resources to historians and sociologists of science more generally. It suggests the possibility of fruitful research on just how young women become adult persons in the context of other kinds of career aspirations. How do young women negotiate their gender identities in classrooms, other training programs, and on the job as they seek to become philosophers, electricians, artists, airline pilots, surgeons, and business executives? And how do young men negotiate their gender identities as they seek to become nurses, ballet dancers, and kindergarten teachers?

Meanwhile, there is reason to rejoice that equity efforts have paid off, at least in many science classrooms, though little such success is yet visible in most fields of professional science education.[30] Moreover, equity is just one of the goals of feminist work in science education.[31] For example, in a more comprehensive sense of the term, feminists have drawn attention to the "scientific illiteracy" of elites about the unacknowledged sexist and androcentric projects visible in scientific research and its various institutions, from science and technology museums to *National Geographic* and *Discovery* TV programs. The natural sciences and their publics have lacked the critical resources to enable them to think about the sciences' complicity with the antidemocratic gender, racist, national, colonial, and imperial projects of their societies. In this respect they have been only incomplete or semisciences, an issue to which I return in later chapters.

GENDER AND TECHNOLOGY

From the beginning of the women's movement in the 1970s, there were projects intended to gain access for women to technological skills and practices from which they had been excluded. For example, courses on car maintenance and on repairing household appliances were offered through the new women's centers. Women were encouraged to enter male territories in the construction trades and the emerging information technologies, as well as agriculture and engineering schools. Accounts of women inventors appeared. Feminists entered the "appropriate technology" movement and struggled to draw attention to feminist analyses of what constitutes an appropriate laundry, cooking, or health care technology.

But it took the arrival of social constructivist analyses in technology studies to open the way for deeper understandings of how technologies themselves were gendered. First, the object of study shifted from the mechanical nature of the "hardware," the technological artifact itself, to the processes of technological change. Such processes are always sites of interlocked class, racial, imperial, cultural, and gender struggles. Processes of technological and scientific change participate in the emergence of new social formations. Second, such change was understood to have three components: changes in "hardware," that is, in artifacts; in the skills required to design, use, and repair artifacts; and in the organization of labor with such skills. Who did and who did not get to design, use, and repair cars, washing machines, and computers? Thus, third, explanations of technological change require attention to how class, racial, cultural, and gender projects of the larger social formations often instigate technological change and how struggles against such projects have succeeded or failed to change how technologies will be used. David Noble's study of the historical links between men's fantasies of bodily transcendence and their psychic investment in technological mastery provides a distinctively valuable approach to this topic.[32] Finally, scientific methods and other practices are themselves technologies of knowledge production. In this way social aspects of technological change permeate sciences' cognitive, technical cores.[33] I return to this topic in the next chapter.

Important new analyses of feminist technological issues have appeared in the Third World feminist projects that look at the role of Western science and technology in globalization, a topic to which I return in a later chapter.

* * *

To conclude, feminist science studies have continued to probe in familiar directions how modern Northern sciences and their philosophies discriminate against women. And they have sparked new debates and discussions far beyond the borders of feminist institutions. What about feminist epistemology and philosophy of science? The next chapter takes up this topic.

5

Discriminatory Epistemologies
and Philosophies of Science

What does it mean to say that epistemologies and philosophies of science discriminate against women, poor people, gays and lesbians, and peoples not of European descent? For mainstream epistemologists and philosophers of science and their followers in natural and social sciences, such a claim makes no sense. Legitimate women's issues about the sciences are restricted to concerns about discriminatory social practices, sexism in the results of research, sexist and androcentric science education, and discriminatory applications and technologies of science—the topics of the last chapter. Even some feminists share this view. For most of these researchers and scholars, the claim that epistemologies and philosophies of science discriminate against women is unintelligible; indeed, it appears to them to be absurd. If the methods of the natural sciences, properly used, ensure value-free results of research, as they believe, then surely epistemologies and philosophies of science, which are even more abstract and disembodied than the sciences they have conventionally set out to explain (for they never even mention any actual people), are immune from this sort of criticism.

Feminist research has undermined this assumption in three ways. First, it has revealed how the methods, assumptions, and results of research in "good science," not just in "bad science," have in themselves sometimes advanced sexist and androcentric projects. Conventional standards for objectivity, rationality, and good method appear to be incompetent to detect these sexist and androcentric values and interests, especially (but not only) in biology, medicine, and health, environmental, and social sciences. Thus, responsibility for such lapses cannot be restricted only to individuals or their conscious intentions, since even the research everyone has regarded as meeting those standards has exhibited such cultural values and interests.

Second, it has shown how the purported excellence of those standards is again and again defined in terms of the separation, the distance, of such standards from whatever counts as womanly or feminine. (These standards are also defined against "the primitive," as antiracist and postcolonial accounts have argued.) Scientific standards themselves are always already actively politically engaged, whether or not those who support and follow such standards intend the particular politics that the standards promote. Sciences and their philosophies have a "political unconscious" (as will be discussed in chapter 7).[1] Thus, merely literal, explicit value neutrality, such as the avoidance of overt sexist and androcentric beliefs and behaviors, is insufficient to block cultural influences on the standards of science.

Third, feminist research has demonstrated that it is only those politically engaged sciences and their philosophies that insist on the responsibility, the accountability, of the sciences for their consequences, intended or not, that can gather the resources to detect such discriminatory assumptions and practices. Neither the scientific nor the moral responsibility of scientific institutions and practices can end at their official borders with life outside the laboratory (so to speak). The consequences of scientific work do reflect back on the scientific adequacy of that work as well as on its (intentional or not) moral and political commitments. Moreover, not all political and cultural influences can be said to block the growth of knowledge; evidently, some can advance it.

However, as indicated, not all feminists—or even all feminist science studies scholars—have agreed with these kinds of claims. For these researchers, there is nothing wrong with the conventional standards; they just have not been rigorously enough followed or monitored. Sometimes these feminists blame sexist scientists—individual "rotten apples" in the otherwise exemplary scientific community—for such "bad science." Moreover, some Third World feminists point out that in their kinds of societies, many women as well as men see modern science as still promising liberation from fundamentalist religious beliefs that both block the production of knowledge about nature and society and oppress women and other groups. (I set aside until the next chapter this important issue of the different position modern sciences and their technologies and philosophies occupy in different cultures.) To be sure, even in the most modern contexts of the metropolitan centers in the West, the sciences still must struggle against age-old "biases and superstitions" that sometimes retain wide public support. Christian "creationism" histories of human development are just one of the most visible such traditional resistances to modern science.

Yet an array of critics has insisted that the very standards of good science and the institutional practices that support them must bear responsibility for the systematic production of cultural values and interests in the practices and results of scientific research.[2] Philosophies of scientific knowledge turn out to be no more immune from cultural influences than are the patterns of the claims of biology, physics, chemistry, sociology, economics, and psychology that they try to explain. Sciences participate in the gender relations of their historical era. Thus, some feminists have taken on the project of revising and strengthening standards for the objectivity, rationality, and good methods of the sciences.[3] This project already distinguishes these feminists from contemporary sociologists and philosophers of science, who show little or no interest in intervening in the practices of science in order to enable the sciences better to serve feminist and other social justice political goals.

Feminist attention to epistemologies and philosophies of science has entered a new phase in the past few years. One focus of such attention has been the development of a deepening understanding of the implications of feminist standpoint theory for research practices, not just for abstract philosophies of science, and of the reasons for its continuing philosophic and methodological controversialism in the face of its continuing dissemination. Another, related, focus, sometimes supportive and sometimes critical of standpoint theory, calls for the abandonment of "representationalist" theories of the nature of science and epistemological conceptions of the philosophy of science in favor of a focus on scientific practices. Scientific practices inevitably must "deliver" nature to us in an already discursively encultured form. In both cases feminist work has moved past the mainstream philosophy of science and science studies.

Standpoint Research: Its Methodological Nature

In order to identify new understandings of standpoint research, we must briefly begin at the beginning—or at least at the beginning of its development by feminists. Feminist standpoint theory emerged completely independently in the work of sociologist Dorothy Smith, political philosopher Nancy Hartsock, and sociologist of science Hilary Rose in the 1970s and early 1980s.[4] They reconfigured the logic of the older Marxian "standpoint of the proletariat" to serve the feminist needs that the prevailing empiricist and interpretative approaches to social science could not provide. This work was quickly joined by that of scholars from a number of disciplines, including from the philosophy and history of the natural sciences.[5] Consequently, it was developed in disciplinarily diverse directions. It appeared as a theory of knowledge, a general philosophy of science, a sociology of science, a methodology, and an empirical explanation of the surprising successes of the early feminist natural and social science work. It was also perceived by these feminists as itself a form of scientific and emancipatory political practice, and as a guide to how to make the practices of natural and social science research, not just their products, serve liberatory goals as well as more ambitious scientific ones.

CENTRAL THEMES

In spite of the diverse and sometimes conflicting standpoint claims of these researchers coming from different disciplinary conditions, such accounts do share important features.[6] I stress here the effect of these features on research methodology, on theories of scientific practice, since I think it is its value to research practice that accounts for standpoint theory's continuing dissemination in the face of continuing theoretical criticism.

First, a theoretical claim. Standpoint approaches argue that all knowledge is "situated knowledge." It is constituted in historically distinctive ways—for example, as part of androcentric and Eurocentric cultural projects or as part of feminist and antiracist projects. These approaches to research reject the possibility of anyone being able to speak from no particular social or historical location at all—what Donna Haraway has characterized as the "God trick" of modern philosophies of science.[7]

Second, standpoint approaches intend to, and can, produce research that is "for women." Such research practices answer questions that arise from women's lives and are about the dominant social order, questions that cannot be asked within the prevailing conceptual frameworks. For example, they ask questions about who benefits from women's double day of work, who controls women's sexuality, and why objectivity, rationality, and good method are persistently linked to certain models of masculinity, and only dominant Western ones at that.

Thus, third, standpoint research is by its very focus overtly politically engaged. Of course, all research is "socially situated," and thus engaged, intentionally or not, in the particular social, economic, and political assumptions of its era. Standpoint practices are overtly politically engaged in their conscious, intentional critical focus on the power relations that oppress women and other economically and politically vulnerable groups. Such research announces its accountability, its responsibility, both to "the facts" of nature and social relations and to ending the oppression and exploitation that have been maintained in part through suppressing or neglecting to notice such "facts." Standpoint projects do not substitute politics for purportedly value-neutral scientific standards; the latter do not exist. They substitute feminist concerns for oppressive and scientifically obscuring values and interests that, for example, have shaped sciences' standards for objectivity, rationality, and good method. Philosophy of science, epistemology, methodology, and scientific research projects are always also political ones, whether or not their agents recognize the political values and interests of their work.

Fourth, standpoint approaches "study up": they study dominant institutions and their practices and cultures, rather than focusing only on groups less powerful than the researchers themselves, as is characteristic of much social science research, or only on purportedly culture-free nature, as is characteristic of the natural sciences. Standpoint approaches focus on the "conceptual practices of power," in Dorothy Smith's felicitous phrase.[8] They identify ways that the research disciplines are complicit with power as they work up the confusing experiences of daily life into the categories and causal relations between such categories that will permit women and other politically vulnerable groups to be governed by elites, as Smith puts the point. They are by no means restricted to recording "women's voices" or the voices of other marginalized groups, important as such voices are. Nor are they restricted to providing ethnographic descriptions of women's lives, important as such projects can be. Their objects of study are the dominant institutions and men's lives, and the effects these have on women's lives.

Fifth, research that would identify the conceptual practices of power must start off researchers' thought from women's lives instead of from the disciplinary or social policy conceptual frameworks that justify, often by treating as natural, women's oppression, domination, and exploitation. This directive has been understood in three ways: to start off from women's experiences, from what women say; to start off from women's objective social locations, as sociologists or economists would chart them; and to start off from feminist discourses, from the social movement analyses and manifestos, legal arguments, and scholarly accounts that have been produced in the name of feminisms or on behalf of groups of women. Each way of starting off research from women's lives can be useful; no one is sufficient. Nor can there be any essentialized "woman's life" that standpoint approaches mark as the origin of feminist research. Just which women's lives one could best start off from depends on the focus of the research project.

Yet, sixth, whatever strategy a researcher may use to start off inquiry from women's lives, a standpoint is an achieved and collective position, not an ascribed position or an individual opinion. A standpoint requires "science and politics," as Hartsock puts the point: "Because the ruling group controls the means of mental as well as physical production, the production of ideals as well as goods, the standpoint of the oppressed represents an achievement both of science (analysis) and of political struggle on the basis of which this analysis can be conducted."[9]

Finally, standpoint-directed research is able to practice more effective methodological strategies and thus produce more objective accounts of nature and social relations than conventional research that attempts to achieve value neutrality. It is committed to maximizing "strong objectivity," rather than only the weaker objectivism that permitted so much sexism and androcentrism to escape detection.[10] Mainstream objectivism's value neutrality is an impossible goal to realize, since we can never completely transcend the conceptual constraints of our particular historical moment. And it is an inappropriate goal because it blocks trying to understand which values and interests in practice do advance the growth of knowledge. Indeed, as a methodology—or a method, as both Smith and Hartsock assert—standpoint theory directs the critical identification of kinds of values and interests in research practices that the conventional standards for objectivity and "good method" could not do. And it raises the challenge of identifying which values and interests have which effects on the production of knowledge. What are the social, economic, and political situations and commitments that can, at least in principle, produce scientific advantage?

WHAT'S NEW?

These standpoint themes were already well developed in the initial standpoint writings identified above, though poorly understood by many observers. What is new about how this approach is understood today?

Standpoint theory is even more widely controversial today than it was initially. Yet at the same time it has also been widely disseminated and is used in virtually every social science and in several fields of the natural sciences. Its critics (and even fans) charge it with various conflicting positions: rejecting modern science or still loyal to its assumptions or both, modernist or postmodern or both, relativist or absolutist or both, relevant to only the social sciences or finding its greatest achievement in its critique of conventional natural sciences and their philosophies, essentialist and Eurocentric or adaptable to any kind of knowledge seeking in the context of inequality, and against disciplinary conceptual frameworks yet preoccupied with debates within particular disciplines.[11] As philosopher Alison Wylie points out, "Standpoint theory may rank as one of the most controversial theories to have been proposed and debated in the twenty-five to thirty year history of second wave feminist thinking about knowledge and science. Its advocates as much as its critics disagree vehemently about its parentage, its status as a theory, and, crucially, its relevance to current feminist thinking about knowledge."[12] Three decades is a long time for a theory to remain vividly controversial. What accounts for this long history and its particular sources of controversiality?

One possibility is that its controversialism itself is a symptom of its desirably innovative features, and in particular of the way it refuses to engage in a number of traditional debates, instead bypassing them to negotiate conceptual and practical commitments that work best for its liberatory goals.[13]

Consider, for example, its complex relations to disciplinarity, upon which the structure of universities and of research funding is based. Here its controversialism is linked to concerns about scientifically and politically appropriate research practices. For one thing, standpoint approaches are multidisciplinary such that researchers with different disciplinary interests have made distinctive contributions to the way it is generally understood. For example, sociologists of knowledge, political philosophers, and philosophers of science have contributed to it distinctive resources useful to whomsoever uses it to direct or justify research projects. Yet it is also transdisciplinary, providing a model or plan for how to do liberatory research that can and does function in every discipline that participates in manufacturing "conceptual practices of power." These two features have led critics to disclaim its relevance to their own particular discipline: it is only a sociology, the philosophers say, and therefore irrelevant to the philosophy of science; it is only an epistemology, the sociologists respond, and therefore lacks the empirical grounding in research practice that any claim to knowledge should be able to exhibit. Yet the standpoint arguments show how dependent sociological and philosophic assumptions are on each other, contrary to their conventional stance of purported autonomy. Moreover, the plot thickens since standpoint practice is also antidisciplinary, focusing on the distorting power of disciplinary loyalties and the complicities between disciplinary projects and conceptual practices of power. Yet, paradoxically, it is also deeply disciplinary, as those who articulate it mostly do so in the context of the conceptual practices of a particular discipline. Consequently, it has developed in different ways as it has engaged with the prevailing conceptual frameworks of the sociology of knowledge, political philosophy, or philosophy of science, for example.[14]

The plot thickens even more. Standpoint practices are both scientifically and politically engaged. They are not "disengaged" either from how research is practiced or from political goals. Thus, beyond the various relations to disciplinarity in which standpoint practices participate, they are openly intended to change how research is done and the political ends that research serves. Moreover, standpoint advocates argue that these political commitments in fact improve the scientific quality of research. Existing research is so deeply complicitous with the conceptual practices of power that only conceptual practices overtly opposed to the currently reigning forms of power stand a chance of producing those accounts of reality that remain unintelligible or illegitimate for the prevailing ways of thinking. Accounts of who actually bears the costs and who receives the benefits of existing governmental and disciplinary practices turn out to conflict with the purportedly democratic goals of governments and disciplines.

Another reason for the controversialism of standpoint projects is that they take on philosophies of the natural sciences whether the particular research project is focused on natural or social science research. We are used to criticisms of insufficiently rigorous practices in the sciences. But standpoint projects challenge the adequacy in their own terms of standards for good research. These standards have not been competent at detecting how certain kinds of powerful cultural values and interests—androcentrism, white supremacy, Eurocentrism, exploitative class interests, imperialism—shape research projects, yet this was supposed to be the task of these standards. Thus, standpoint projects challenge familiar ways of thinking about ideals of modernity, democracy, and social progress, ideals each dependent on the adequacy of modern Western scientific research standards. These ideals have helped to constitute individuals' and social institutions' identities and widespread conceptions of what are legitimate and important social projects. They direct practices. Standpoint theorists, in contrast to their critics, have largely neither exalted nor rejected such ideals but rather have tried to refashion them so they can function to justify and guide more democratic practices.

Perhaps one of the most controversial aspects of standpoint research is its conception of the subject of knowledge as collective. It rejects epistemological individualism.[15] Of course, other feminist epistemologists and science theorists in general point out that whatever gets to count as knowledge must have been checked and rechecked, articulated and discussed, by scientific communities. Scientific communities are the actual subjects, "authors," of scientific claims, and so such communities should not themselves discriminate on the basis of gender, race, ethnicity, sexuality, class, or other such social (or biosocial) markers.[16] This is certainly a worthy goal. Yet standpoint approaches reject epistemological individualism in an additional way. For standpoint approaches, it is the collective experience of subjugated groups, when and as it can be articulated as such, that in itself constitutes those groups collectively as the subjects, the "speakers," of research that is *for them*. Such collectively articulated experiences bring such groups into existence for the group (as opposed to "objectively," for others who are mapping their beliefs and behaviors). Standpoint practices of "studying up" to identify the causes of a group's own immiserated conditions contribute to the creation of such liberatory consciousnesses.

Of course, women in different classes, races, cultures, and historical contexts, and subcontexts within any such category, have very different experiences. Standpoint theory is not committed to essentializing women; indeed, its logic is precisely opposed to essentializing practices (which is not to say that standpoint theorists never fall into an essentializing moment). Though the practices of slavery are objectively diverse, and are experienced differently by women and men, domestic and field workers, those with kind or cruel masters, there is nevertheless something about the slave experience in itself that creates the possibility of articulating the conceptual practices of power that justify and maintain such an institution. Thus, standpoint approaches enable groups' collective experiences of domination, exploitation, oppression, and discrimination to be turned into distinctive resources for knowledge and for political liberation. They enable new kinds of questions and debates about the role of experience in the production of knowledge. Fredric Jameson argues that it is feminist standpoint theory's particular kind of challenge to the philosophy of the natural sciences, and its development of a new kind of collective consciousness for women as agents, subjects, of knowledge, that has enabled it to open up the topic of the role of historical social experience in the production of scientific knowledge in a new way that Georg Lukacs and other Marxian science and knowledge theorists desired but could not achieve.

> Today, one has the feeling that the most authentic descendency of Lukacs's thinking is to be found, not among the Marxists, but within a certain feminism, where the unique conceptual move of *History and Class Consciousness* has been appropriated for a whole program, now renamed (after Lukacs's own usage) *standpoint theory*. . . . These path-breaking texts now allow us to return to Lukacs's argument in a new way, which opens a space of a different kind for polemics about the epistemological priority of the experience of various groups or collectivities.[17]

Finally, a number of innovative moves central to standpoint approaches from the very beginning have become the object of new scrutiny and widespread debate.[18] Some of these have been identified above. What does it mean to "study up," to identify and scrutinize the conceptual practices of power? How is it possible for standpoint theory to be both a modern and a postmodern theory, as different groups of critics have charged? How is it possible for standpoint theory to escape a damaging relativism? What is the "reflexivity" standpoint theories advance and claim for themselves? How is it possible for a research project to engage in politics and still assert greater objectivity for its claims? Is it really committed to combatting class oppression when it departs from so much of the Marxian conceptual and political framework? What is the relation between the standpoint claims articulated in the early writings of Dorothy Smith, Nancy Hartsock, Alison Jaggar, Hilary Rose, myself, and others and the approaches of other oppressed groups, whether or not these are so articulated as standpoint practices?

Paradoxically, the controversialism of standpoint theories turns out to be a resource. Defending this point is itself controversial since the usual goal of theorists is to make their own position uniquely compelling. Certainly that, too, is a goal of standpoint theorists. Yet since there is disagreement among standpoint theorists, coming as they do from different disciplinary and even political projects, I am arguing that we should understand this theory's controversialism in a different way. Reflecting on standpoint theories and the research projects that they direct provides useful sites for discussion and debate over how the disciplinary structure of research institutions should be reorganized and how we should transform conceptions of modernity, democracy, progress, and, indeed, science itself. Transforming our conception of the fundamental nature of both science and the nature it studies is the project of a related focus in feminist philosophy of science.

How Scientific Practices Produce Encultured Nature

Philosophies of science have always understood how "bad sciences" project onto nature cultural and political fears and desires. To such cataclysmic natural forces as tornadoes, hurricanes, snowstorms, droughts, floods, earthquakes, and volcanoes have been attributed malevolent intentions toward humans. Or to men are attributed uncontrollable sexual needs and to women "inherent" needs to bear and take care of children. Such sciences have been charged with failing sufficiently to distinguish nature's own properties from those feared or desired in particular historical communities.

Yet the past few decades of science studies have shown how cultural fears and desires inevitably influence research projects. For example, feminists have pointed out how culturewide sexist and androcentric fears and desires have shaped the objects, conditions, and processes on which sciences focus; what is considered problematic about them; the hypotheses favored; the concepts, metaphors, and models of nature featured in such hypotheses; the design of research projects; the way data are collected and interpreted; decisions about when to end experiments; and how the results of research are articulated and disseminated. Sciences inevitably participate in their historical era. It would seem that no more need or could be done to demonstrate that and how scientific projects produce encultured results of research.

Yet recent work by feminist philosophers of science, Karen Barad and Joseph Rouse most notably, has developed in newly compelling detail this kind of analysis by focusing on how scientific practices themselves necessarily enculture the nature that appears in the results of research. The linguistic and conceptual practices of the sciences—scientific representations of nature—had been the main focus of conventional philosophies of science. Following recent tendencies in science studies more generally, Rouse and Barad have insisted on the importance of shifting philosophic attention from sciences' representations of nature to scientific practices or interactions with nature more generally. On this kind of account, representations of nature are just one of many kinds of scientific practice. This account rejects what it calls philosophy of science's "epistemological" project of surveying and justifying a domain of scientific knowledge imagined as bounded, that is, with fixed borders. It provides a different account of how nature's "matter" presents itself to scientists as already culturally, politically constituted yet still does powerfully participate in producing scientific claims. This feat simultaneously counters both the conventional tendency to regard matter as anormative and the postmodern tendency to focus only on the normative linguistic and conceptual aspects of science. From this work has come another stronger notion of objectivity than prevailing value-neutral ones. This work shows how culture and politics routinely enter scientific labs and then escape into the everyday world through scientific practices. Finally, it reveals the importance of a politics of responsibility and accountability with respect to scientific practices that had been presumed to be sufficiently monitored by research methods for eliminating social values and interests from research. I shall argue that these two kinds of work—standpoint research and this second kind of philosophy focused on scientific practice—complement each other: each provides analyses that the other needs.

AGAINST REPRESENTATIONALISM AND
PHILOSOPHY OF SCIENCE AS EPISTEMOLOGY

Doubts about the representationalist conception of science are not new. Thomas Kuhn argued in *The Structure of Scientific Revolutions* that innovative scientific practices have preceded their theorization. For Kuhn, a paradigm is a particular way of conducting research, a set of practices, including linguistic and conceptual ones, on the basis of which normative principles to guide practitioners could only later be formulated. Michel Foucault described in *Discipline and Punish* how the practices of new institutions of social control in the nineteenth century, the prison and the clinic in particular, prepared the way for the emergence of the social sciences. The particular practices of observing while disciplining bodies that these institutions introduced provided a model for emerging social sciences for how to systematically gather information about people. In the 1980s, philosopher of science Ian Hacking argued in *Representing and Intervening* that conventional philosophies of science overemphasized sciences' productions of representations of nature's order at the expense of their more central concern with successfully intervening in nature's processes. In these accounts science is reconceptualized as fundamentally about achieving successful practices rather than producing accurate representations of nature's order—"truths." To be sure, representing nature is one kind of scientific practice, but it is just one among others.

Since the 1950s, philosophies of science have shifted their focus from earlier attempts to produce "rational reconstructions" of an idealized logic of scientific inquiry, such as the hypothetico-deductive model, to philosophic accounts of the actual practices of scientific research, regardless of whether they match some idealized logic. They have been assisted by the outpouring of histories, sociologies, and ethnographies of actual scientific projects and practices.[19] In shifting their focus in this way, such philosophies can appear to have turned away from conventional epistemological preoccupations with surveying and justifying scientific representations of nature. They can appear to have abandoned representationalism in favor of practice-focused philosophies.

Yet the assumptions of representationalism and conventional epistemologies have not been as fully abandoned as this legacy would appear to require, Rouse has argued.[20] Epistemology assumes that there is a bounded, coherent, in principle knowable collection of representations of nature's order that it is science's task to identify. Social constructivist arguments consistently undermine such an assumption. The total of possible illuminating representations of nature is infinitely large, as historically new cultural preoccupations will continue to call for and produce new ways of understanding ourselves and the changing world around us. Indeed, multicultural and postcolonial science studies make the "epistemological" assumption appear absurd.[21]

Moreover, philosophers should not be wasting time justifying scientific claims at all, Rouse argues; all the explanation and justification needed are already provided by the sciences themselves. Moreover, there are important issues philosophies of science need to address, such as how culture and politics can advance the growth of knowledge and how they enter nonlinguistic scientific practices, that do not lie within the conventional epistemological concern with justifying science's knowledge claims. Most problematic for social justice movements is that epistemology's representationalist understanding of scientific knowledge makes it difficult to understand relations between knowledge and power.

Rouse's own early work argued against such an epistemological and representationalist understanding of science. However, he points out that feminist science studies has made distinctive contributions to the development of such a position. He argues that this is evident in feminists' focus on the relation between knowers and the known instead of only on the content of knowledge—that is, on representations of nature and social relations. It appears in feminism's participatory stance toward science rather than an interest to explain or assess it as a totality. Feminists are concerned for the future of science rather than only about the present state of knowledge. It is apparent in a refusal to distinguish between epistemological and political projects (or to reduce one to the other), and in a commitment to a notion of reflexivity that requires that other voices be heard for their own sake, not just to satisfy sociologists.

One can add to Rouse's account here that for feminist researchers, a robust reflexivity practice requires not just listening to other voices but also trying overtly to position one's own work in contemporary social relations in the same ways researchers position the beliefs and actions of those they study. That is, robust reflexivity requires articulating how "this very account," the one I am producing here and now, is made possible and desirable by social and political conditions. This robust notion requires doing the work of analyzing the consequences of others' comments and criticisms for one's own work rather than only "tolerantly" listening to and perhaps reporting them.[22]

Rouse argues that the research practices of the natural sciences do change the world, contrary to conventional claims that science leaves the world it studies as it finds it. For example, our medicine cabinets and kitchens replicate laboratory conditions; we discipline our bodies to carefully reproduce the laboratory temperature, humidity, careful measurement, isolation of substances from each other, and appropriate manner of handling them. Lab materials and practices "escape" the lab into our daily health and food preparation routines.[23] Thus, Rouse shows how the kinds of organizations of nature that scientific practice itself requires, whether in the lab or in the field, cannot possibly be culturally neutral.

How Matter Matters to Science:
Performative Scientific Practice

More recently, in a series of essays and a forthcoming book, physicist and philosopher Karen Barad has taken on the ambitious task of providing a conception of nature that better fits the goals of feminist science studies, and feminist philosophy of science in particular.[24] Barad draws on Rouse's work, as his work in turn has drawn on hers. In effect, Barad, like Rouse, shows that philosophies of science that cannot recognize how culture and science coconstitute each other through scientific practices end up discriminating against those groups disadvantaged by the values and interests that scientific practices do in fact produce. Such values and interests shape the conception of "nature" that appears in the results of scientific work. Barad's project also rejects both representationalism and the epistemological conceptions of science and the philosophy of science in favor of a focus on scientific practices. For Barad, scientific practice reconfigures the world such that it appears as already conceptually articulated and as having normative consequences. Thus, politics flow from scientific practices seemingly without human intent.

Central to Barad's account is the notion of a "phenomenon." Her example is an experimental setup—the measuring equipment, the scientists, the segment of nature to be measured, the particular socially situated laboratory and its larger social context. Yet her argument here can be extended to any other kind of situation into which a scientist enters. Such a phenomenon is always already both material and discursive, and therefore already normative. Scientific description (in its classical form) requires measurement, and measurement requires discursive practices that would permit an appropriate description. Knowers, or agents of scientific knowledge, are coconstituted with possible ontologies of the world in what Barad calls "intra-actions" with phenomena. Such intra-actions allow the world to be intelligible. Science reveals the world as already normatively constituted. Barad draws here on philosophers J. L. Austin's and Judith Butler's notion of "performativity" to argue that it is these intra-actions that simultaneously bring into existence an already normatively meaningful and, thus, "agential" world and human knowers. For Barad, maximizing objectivity consists of maximizing agential responsibility.

Barad is concerned to oppose both the conventional image of anormative nature that can passively be reflected in scientific representations and the postmodern image of a reality completely constituted by language or discourses, with apparently no causal role for prediscursive matter. Her ontology enables us to see how the material and discursive experimental setup to which Rouse refers does actively bring into existence our kitchen and medicine cabinet practices as well as us as knowing agents. Sciences' setups also bring into existence normative directives about gender, class, racial, or sexuality differences. Nature does play an active role in shaping these differences, but it can do so only when it is already constituted as intelligible through cultures' discursive observational and measurement assumptions and practices.

POLITICAL VALUES CAN BE SUPPORTED BY EMPIRICAL EVIDENCE

One more very recent criticism of epistemological and representationalist philosophies of science deserves at least mention here. Sharyn Clough draws on antirepresentationalist themes in the work of philosophers Nelson Goodman, Donald Davidson, and Richard Rorty, as well as on Rouse's work, to "naturalize" feminist claims about relations between knowledge and power. She deploys Davidson's criticisms of the dualism of conceptual schemes and empirical claims against feminist philosophers of science. "On Davidson's nonrepresentationalist view, our explanatory models, values, political commitments, or worldviews are best conceived, not as conceptual schemes through which the evidence from the external world is filtered; not as underdetermined representations of the evidence; but as further strands in our web of belief." By giving up this kind of dualism, feminist science studies can fully defend its worldviews and political values without committing itself to general epistemological standards and their representationalist assumptions. "When we justify particularly critical elements of our feminist worldviews, such as our beliefs about oppression and justice, our appeals to (beliefs about) the empirical data have been well documented and are powerfully persuasive as a result. There is no need for us to doubt the empirical evidence for our feminist political values, as long as we conceive of the evidence as that which is provided by other beliefs in our web." This is a different way of conceptualizing the relationship between scientific and political claims: a political claim is simply one more for which the evidence can be evaluated in ways no different from how one evaluates the claims of some particular scientific project.[25]

The Importance of Feminist Practice-Focused Philosophies of Science

The innovative accounts highlighted in the preceding section correct for problematic tendencies in contemporary philosophy, history, and sociology of science, including feminist work in these fields. These examples of practice-focused philosophy of science are much more complex and interesting than this brief focus on them can convey. I hope that readers now have enough sense of their main themes, however, to enable us to turn to some reflections on their collective significance and their relation to standpoint theories.

First, as indicated earlier, sciences and their philosophies that cannot recognize and do not engage with how scientific practices themselves inadvertently legitimate and further disseminate political and cultural values and interests usually end up complicitous with the agendas of dominant social groups. These sciences and their philosophies usually discriminate against the most politically and economically disadvantaged groups no less than more overt forms of discrimination. I would say they are epistemologically underdeveloped had that "*E* word" not fallen into disrepute in this chapter (but see below).

Second, Rouse argues that feminist work differs from the sociology of scientific knowledge that, with the new histories of science, has provided influential directions in post-Kuhnian science studies. The feminist work opposes relativism. It takes a normative stance to particular scientific claims. It is willing to revise standards for objectivity, evidence, and the distinction between belief and knowledge. And it has developed a postepistemological conception of knowledge (as that term was developed in this chapter following Rouse), in contrast to both the traditional philosophy of science and the sociology of scientific knowledge. Rouse's argument here clarifies why it is that the sociology of scientific knowledge, which has in other ways informed much feminist work, has not returned the compliment to feminist philosophy of science: it remains largely uninterested in feminist science studies. To put Rouse's point here in a slightly different way, the sociologists apparently see the feminist work as beyond the borders of the intelligible precisely because feminism rejects the features of the epistemological conception of knowledge that leave the sociologists in league with just the mainstream philosophies of science from which the sociologists take themselves to have made a decisive break.

Third, feminist standpoint projects, along with other feminist science studies work, have from the beginning been interested in scientific practices and how to improve them, as Rouse points out by including them in his list of innovative feminist science studies. Yet they have also been concerned to provide justifications of feminist knowledge claims—a practice that the antiepistemology and antirepresentationalist projects would delegitimate. I think there remains important work for this particular kind of epistemological project. Feminists have been using that term in ways unintelligible to conventional epistemologists, as Rouse in effect argues.[26] Yet they also need to justify feminist empirical work in ways that go beyond the evidence such work itself provides. This work is consistently ignored or disputed, as observers have again and again pointed out. "The facts" do not announce their own legitimacy, evidently, when they challenge the favored assumptions of dominant institutions. "The truth" does not "set us free" all by itself. Epistemological claims of the standpoint sort provide another kind of empirical evidence for taking seriously certain classes of "facts" about nature and social relations. "Start out thought from the lives of oppressed groups and you will gain different insights about the dominant social institutions, their practices and cultures, and the nature and social relations that they 'rule,' than you can gain by starting off thought from the favored problematics of those institutions." Such claims are not free of culture and politics, but then, as Barad and Rouse argue, neither are the scientific "facts" they support. Standpoint epistemologies are a kind of "science of science," one could say. Perhaps it is just confusing to refer to them as epistemologies at all; perhaps they should indeed be called only sociologies of knowledge. Yet they came into the world as empirical and normative social theories of what does and should count as knowing. It is the "should" that leaves them, I am arguing, in the realm of epistemology's normativity. The feminist refusal to separate epistemology from politics creates this odd situation, I suggest.

Fourth, what these antirepresentationalist, postepistemological accounts do not do is identify just which scientific practices will in fact advance prodemocratic social justice responsibility and accountability. They are completely silent on such matters. This is the point at which one can see that these approaches are in fact complementary to standpoint projects, which try to identify which kinds of scientific practices do take on the responsibility of their political accountability. Standpoint approaches try to remove "good scientific practices" from the realm of the voluntary; they provide a specific normative principle to guide scientific practices.

Finally, the achievements of these two recent tendencies—especially the practice-focus philosophies of science—have not prevented them from remaining Eurocentric and "part of the problem" for the majority of the world's citizens to the extent that they fail to engage substantively with critical perspectives provided by studies of the high integration of Western sciences into profiteering and militarism, and with multicultural and postcolonial science and technology studies, which were addressed in earlier chapters and will be again in the next chapter.

6

Feminist Science and Technology Studies at the Periphery of the Enlightenment

Women in the South want from sciences much that women in the North value. They want improved access to appropriate science and technology education and to the kinds of income-producing work such training makes possible. They want better access to effective health care and workplace technologies. They want safe and flourishing natural environments. They want a significant voice in the science and technology decisions that affect their lives—decisions made primarily in science labs and in international and national agencies.[1] Moreover, Western modern sciences and technologies are linked to desires in the South, too, for modernity, democracy, and social progress. For many citizens of the South, as in the North, to get to participate in modern international sciences and engineering is to leave social environments concerned with only parochial interests and to enter global conversations, to become citizens of the world. Often Western modern scientific methods and the facts that they produce are experienced as a welcome alternative to traditional discriminatory and just plain ineffectual local beliefs and practices.[2]

Yet perceptions of science and technology are as conflicted and polemical in the Global South as in the Global North. Many debates and discussions of the topic, including feminist work, occur in the context of the science and technology movements that emerged in the South after World War II. These movements started their analyses from the lives of poor people in the Third World. They took their questions, conceptual frameworks, and practices from the periphery of the Enlightenment. Feminist work in the context of these movements has raised some issues that are different from those at the center of Northern feminist concerns. Neither the category "women" nor "feminism" is homogenous. Insofar as women, like their brothers, live in different material and social conditions, different groups of women and of feminists will have distinctive questions and concerns about scientific and technological theory and practice.

Feminism and the Southern Postcolonial Science and Technology Movements

THE COMPARATIVE ETHNOSCIENCE MOVEMENT

This movement has drawn attention to ways in which women have a particular standpoint on nature wherever women and men are assigned different interactions with their own bodies and with the natural and social worlds around them.[3] For these purposes, it can be illuminating to treat gender differences as something like cultural differences in a local environment of inequality. Thus, in particular historical contexts women must interact with local natural and social environments (though nature and culture are never as separable or distinct as such a formulation suggests). Consider, for example, the different interests of women concerned with the relation between apparent increases in cancers and living "downstream" from toxic industries and, in contrast, tribal or peasant women living on the edge of the expanding Sahara desert, who experience decreasing supplies of water, food, and fuel, which they must supply to their communities. These two groups of women have gender- as well as class- and culture-distinctive interests in the world around them, leading them to ask at least a partially different array of questions than appear important to their brothers or to the dominant institutions, and different questions from each other.

Furthermore, women tend to develop their own discursive resources—narratives, models, metaphors, analogies—that, like the discursive practices of men, both enable and limit their interactions with their environments. Moreover, such interactions tend to be organized in the kinds of ways characteristic of women's activities in their class and culture more generally. Finally, to the extent that women are oppressed, dominated, and exploited, they can learn how to organize and collectively call on the distinctive "resources of the oppressed"—on standpoints (discussed in the last chapter)—that give them access to perceptions and interests unavailable to dominant groups.

It is important to recognize such differences while also keeping in focus the value of women finding ways in which their different interests can nevertheless support shared political goals, such as participation in the scientific and technological decisions that affect their lives. Moreover, in many contexts in any culture women and men share common natural and social conditions, activities, interests, discourses, and yearnings. Yet wherever any distinctive gender experiences appear, so too will distinctive questions about nature and social relations and distinctive repositories of knowledge.

Thus, women collectively, in their culturally particular social situations, become repositories of systematic and effective knowledge about nature and social relations. This knowledge is dynamic; like scientific knowledge more generally, it must constantly be revised as women's natural and social environments change. Deserts expand or contract, farmland erodes or is replenished, toxins permeate water and food supplies, new diseases spread or disappear, new ideas arrive on television or from culturally new neighbors or international agencies. And women organize to try to gain access to the resources needed for them and their children, kin, and communities to survive and flourish.

THE "SCIENCE AND EMPIRES" MOVEMENT

The second, partially separate, science and technology movement originating in the Global South that raises women's and feminist issues is the "science and empires" movement. Women's standpoints, or feminist standpoints, also enable gender-illuminated accounts of the causes and nature of the growth of Northern sciences. The "science and empires" accounts are about the mutually supportive relations between the growth of modern sciences in Europe and patterns of European expansion. An earlier chapter identified ways in which each needed the success of the other for its own success. European expansion required new kinds of knowledge if the travelers were to survive, flourish, and dominate each other and the indigenes they encountered. And the development of modern sciences in Europe needed the social and economic support of the sponsors of the "voyages of discovery," as well as the access to far greater natural phenomena that the voyages made possible. Yet these accounts are always also, explicitly or implicitly, about how the Global North's so-called development policies for the "underdeveloped" countries bear an unfortunate continuity with Europe's imperial and colonial policies. Issues about women, gender relations, and sexuality weave through these accounts.

THE SUSTAINABLE DEVELOPMENT MOVEMENT

Development policies, like their forerunners in the days of formal European empire, have benefited primarily the already most advantaged classes in the North. Feminist projects that have started off from women's lives in the South have critically evaluated every aspect of development policies. They have shown how it is the appropriation of women's and peasants' labor and land that has made possible a huge proportion of the increased flow of resources from the South to the North.[4] Of course, these policies were supposed to be raising the standard of living of the "have-nots" in the South to the high level of the "haves" in the North, in the phrases of the day, through the transfer of Northern technologies and their scientific rationality to the South. The South would learn to increase its gross national products through scientific farming, scientific fishing, scientific forest management, and technologically advanced manufacturing, as well as through Western scientific health, medical, educational, financial, and legal practices.

Everyone, or almost everyone, can agree that there is much positive to be said for Northern sciences, technologies, and their distinctive forms of scientific rationality. Yet they have their limits. Their transfer to the South did not prevent development from ending up serving primarily escalating militarism and transnational corporate profiteering while worsening the conditions of precisely the groups in the South that development and its scientific rationality were most supposed to benefit. Peasants, and women and their dependents in other classes, did in the past and do today constitute the overwhelming proportion of the world's most economically and politically vulnerable populations.

This assessment, now regularly available on the front pages of the *Wall Street Journal*, as well as in the reports of the International Monetary Fund and the World Bank that financed many of the problematic development projects, does not require the romanticization of social conditions or of scientific and technological knowledge in the South prior to European expansion. Nor does it call for a return to the past. Rather, it calls for critically re-examining just how it is that Northern sciences and technologies, supposed to advance social welfare for all, have in fact persistently ended up disproportionately providing resources for elites around the world at the expense of the already worst off in the Global North and South. And it calls for creatively envisioning scientific and technological social relations that can effectively block the projects that have such effects.

Two features of these Southern feminist criticisms are important to note. First, central themes in Northern feminist science and technology accounts independently emerged or were importantly transformed in the gender, environment, and sustainable development (GESD) analyses. For example, concerns with the androcentric standards for objectivity, rationality, evidence, good method, and what counts as science were centered in GESD analyses also. Yet new ways of looking at these issues arose not only because such themes ap-

peared in the context of Southern histories but also because the GESD discussion was created mainly in regional and international agencies and in grassroots organizing rather than in the university and laboratory contexts in which
Northern feminist science and technology discussions were shaped.[5] The androcentric structure and meanings of modern scientific and technological
worlds also shaped international, national, and local development agencies and
their thinking. The obscuring and often misogynist dualisms identified by
Northern feminists, postcolonial studies, and the environmental movement
also operated through development rhetoric to shape policies that systematically discriminated against the economically and politically most vulnerable
populations in which women and their dependents are disproportionately represented.[6]

And the questions, the problems, that development policies addressed
were never ones defined by women or from the standpoint of women's lives.
Development was gendered, as even one document from the United Nations
Commission on Science and Technology for Development put the point by
the mid-1990s.[7] Critical examinations of gendered development inevitably
raised questions about not just the applications and technologies of modern
sciences but also their cognitive, technical cores—their ways of conceptualizing nature and inquiry practices.

Second, GESD discussions paralleled and occasionally drew on analyses in
the writings of African American and other feminists of color in the metropolitan centers, which made possible additional coalitions. These discourses,
both in the Global North and in the GESD accounts, helped to redefine subjects of knowledge, the actual or implied knowers or speakers of knowledge
claims, as having multiple and often conflicting identities because of their racial, class, gender, ethnic, sexual, and other histories. Both knowers and their
knowledge systems are "decentered" and heterogeneous in these accounts, and
these features are resources for the production of knowledge. Both in the North
and in the South, these accounts insisted on the importance of empowering
marginalized racial and ethnic groups as a condition of democratic dialogue
and coalition; difference, as well as affinity, must be recognized and respected.
In both the South and the North, these writings have produced a powerful critique of positivism and neopositivisms, and have developed illuminating forms
of feminist standpoint epistemology.[8]

Distinctive Feminist Issues in GESD

Here we can focus on three of the distinctive Southern feminist science and
technology critiques produced in these feminist postcolonial science studies

projects.[9] First, as indicated earlier, these assessments concluded that the cost of development was borne primarily by women and peasants in the South: in its very structure development was gendered. Moreover, second, this "economics" of development was encouraged by the economism of the Enlightenment philosophy of Western modern science, which was thereby especially costly to women and their dependents. Finally, the Enlightenment philosophy legitimates the destruction of the environments upon which Southern women and their dependents rely to stay alive. Modern Western sciences and their philosophies are implicated in each criticism.

IS DEVELOPMENT GENDERED?

A main theme in early feminist criticisms of Northern development policies was that women were being left out of it, as literacy and job-training programs were designed for men only and men were given favored access to the development-sponsored income-generating work. This was so even in matrilineal societies, where women had been traditionally assigned responsibility for the distribution of economic resources. Often the only attention women officially received from development planners occurred under the heading of controlling women's reproduction. Moreover, as men were drawn into "scientific" urban manufacturing, mining or plantation agriculture or lumber production, women were left to become a higher proportion of rural populations, with increased responsibility for the care of the young, old, sick, and disabled and with fewer social and environmental resources to sustain life, the environment, and the community. Development policies bypassed women, the argument went.

Or did they? A second round of analyses showed that these very processes of "leaving women out" were what provided necessary new resources for scientifically modernizing national economies. Achieving economic growth required increasing women's unpaid domestic labor, enticing or forcing women into the lowest-paid manufacturing and agricultural labor (and even into prostitution in the cities where rural men now earned wages), and appropriating their inherited land rights. Their land rights tended to shift to men when only men were taught to farm in modern scientific ways. At other times, these rights were directly appropriated so the land could be used for export production. Women were denied access to community-owned common agricultural, forest, or grazing areas, which were "enclosed"—appropriated for private export production. Peasants as a class, both women and men, also suffered from some of these forms of appropriation.[10]

Thus, the dedevelopment of women and peasants was a necessary condition for development in the South. "Progress for humanity" meant regress for women (and peasants), as Northern feminist historians had put a similar point.[11] Did the scientific and technological rationality being transferred from North to South include encoded "directions," or, at least, permission for such banditry? At any rate, the language of development, modernization, and scientific progress was being used to obscure the actual mechanism responsible for much development success. Structural adjustment policies intended to resolve the debt crisis of the 1980s followed the same pattern, further undermining the conditions of women in the developing countries. Here the International Monetary Fund and the World Bank ordered the indebted Southern governments to cut their social services in order to repay development loans to the Northern investing classes. Thereby, women's unpaid labor in the home was to be substituted for their formerly paid labor in state-provided educational, health, child care, and other social services so as to maintain the wealth of the most advantaged classes in the North.[12] Some of the feminists concerned with these labor issues also began to identify problems with the economism, the production model, of both neoclassical and neo-Marxian economists' attempts to address women's issues, and with the deteriorating environments that development policies created for women's subsistence and wage labor.[13]

IS ENLIGHTENMENT PHILOSOPHY ECONOMISTIC?

Does the Enlightenment scientific ethos depend on a model of human progress as only economic progress? Is the notion of human progress tied too tightly in Enlightenment thought to the ideal of technological advances that can create expanding markets? Of course, this was not the way the great Enlightenment thinkers expressed their ideals. Yet this seems to be the meaning of "human progress" in Western thought claiming the Enlightenment vision at least since the Industrial Revolution. After World War II, development was conceptualized as economic growth to be created by the transfer of Northern sciences and technologies and their models of rationality to the Global South. Thus, human progress was thought of in terms of increased production and consumption.

Feminist critics argued that this approach failed to perceive women's work in the household as real work, or, therefore, as activity that contained elements of a history of human progress.[14] This misconception was as prevalent in the Marxian as in the Liberal analyses. Thus, peasants' and poor people's need for child care and household labor was perceived by development thinkers as a drain on maximum economic growth. Meanwhile, middle- and upper-class women North and South were to be recruited into the consumption work that was required to keep production profitable. Poor women were to be drawn into "productive" agricultural or manufacturing labor and left to get child care and domestic work done as best they could, whereas women in the economically advantaged classes were to devote increased time and energy to purchasing products to enhance child care and domestic tasks, so to consume at higher and higher levels. Thus, to feminists, conceptualizing development and human progress only in economistic production terms left women, the life of the household, and the community life dependent on women's participation intensely vulnerable to exploitation in every class.

This first criticism was related to a second. Modernization theory routinely conceptualized population growth in developing countries as a major obstacle to raising standards of living. Population growth increases poverty, this theory insisted. From this perspective, women's bodies and their sexual activities were major obstacles to social progress, and coercive population-control policies appeared justifiable. However, this kind of account of the relation between poverty and reproduction had the direction of causality backward, it turned out. By the early 1990s, even the United Nations Population Conference officially recognized what feminists and progressive economists had been arguing for years: it is poverty that causes population growth, not the reverse. It takes many family workers, including children, to sustain the daily survival of a family when the work available to them is so labor intensive and their wages are so low. It takes many children to provide poor households with the economic and social supports that the state and the economy provide to middle-class households—unemployment benefits, pensions, sick leave, health care, child care, and elder care, for example. And it takes many more births to produce such child and adult workers when women have poor access to nutrition, prenatal, and children's health care. Conventional Western scientific wisdom had the causal direction backward.

A third problem with the conception of development as economic growth was that nature itself limits economic growth, as feminists, environmentalists, and critics of neoclassical and Marxian economics argued. The world does not have enough natural resources to support today's global population even at the consumption levels of moderately well-off Third World middle classes—classes that would count as lower classes in the metropolitan North. Achieving that standard of living for today's most politically and economically vulnerable populations would require a lowering of consumption levels among the more advantaged portions of the world's population that would be extremely hard to achieve and is virtually unimaginable to precisely the group that must do it. (Think, for example, of the lifestyle changes necessary to get fossil fuel consumption significantly lower in the United States.) What political process could bring this about?

Finally, conceptualizing development in terms of greater economic productivity and consumption ignores and devalues all other "goods" that women and their cultures prioritize, such as ethical, political, aesthetic, and spiritual development. These values are usually seen as obstacles to economic growth, as the restructuring policies revealed again. Fair wages for the worst off, democratic decision making, and the preservation of beautiful urban and rural environments are consistently sacrificed to higher profit returns to investors.

Such considerations lead to the suspicion that there is something deeply irrational about the "rational man," whose interests the dominant economic and political theories consistently advance. This ideal social actor seeks information always in order to maximize his own benefits. Yet he thereby ensures the destruction of the very conditions necessary for his survival when those benefits are conceptualized solely as economic. We should speak, instead, of the irrational economic man idealized in these theories. (This is not a point about individual scientists or institutional actors, but rather about the values and standards of development policies and of other modern institutions concerned with economic development.) Neither nature nor social life can be sustained when such a purely economic rationality provides the dominant institutional logic of states and transnational corporations. Who is to be held accountable for the preservation and development of other social values against the onslaught of such powerful economic forces?

To what extent is this self-destructive rationality an inherent feature of Enlightenment rationality, at least as it is enacted today? Modern sciences emerged as part of European postmedieval economic, political, and social formations. The modern insistence that science must be value neutral, and that knowledge seeking must be protected from any social accountability apart from the standards of "real science," makes modern sciences and technologies "fast guns for hire." Ethics committees in scientific work sites are better than nothing, but they have no power over transnational corporations, which appear unaccountable to any governmental or civic groups at all—even to their own stockholders, as the pandemic of ongoing financial scandals has revealed. Must the sciences internalize more effective and relevant democratic ethical and political principles in order to avoid the "fast gun" status? How are explicit and, even more important, implicit ethical and political commitments of scientific institutions and communities directing scientific and engineering work of the Global North "behind the backs" of purportedly value-free scientific processes? Who benefits and who bears the costs of such ideals as value neutrality, instrumental rationality, and restricted notions of what can count as real science, good scientific method, and the accountability of scientific research and its social institutions? Should the sciences transform their "political unconscious"? What could this mean?[15]

DOES ENLIGHTENMENT PHILOSOPHY LEGITIMATE ENVIRONMENTAL DESTRUCTION?

The growth models of development consistently sacrifice sustainable environments to short-term consumption goals. Moreover, natural resources are disappearing not only through consumption but also through the effects of military activities and agricultural and manufacturing toxic pollution of air, water, and other resources that human and nonhuman life require.[16] Such conditions can discourage progressive groups around the globe who truly want to help bring more equitable social relations into existence. Labor and environmental needs have frequently been perceived to be in conflict in both the labor and the environmental movements. Yet this apparent conflict can also be regarded as a resolvable challenge. One encouraging recent example of such work is the rise of organized protest and shaming of the International Money Fund, World Bank, and global elites and the persistent exposure of their ethically illegitimate policies and practices with respect to both the environment and the world's economically and politically most vulnerable groups.

Women suffer in distinctive ways from the limits that nature places on economic growth, and their disadvantage is passed on to children and others dependent on their energies and resources. They are frequently last in line for economic resources in their households, and disproportionately among the last within their societies. To them is assigned responsibility for doing or managing daily sustenance and the health and welfare of dependents, the household, kin, the elderly, the sick, as well as their communities and environments. Moreover, manufacturing and rural wage labor expose them as well as men to toxic dangers in addition to the toxic threats endemic in poor people's household life, such as vermin, gasses from open hearths, and the like. Life- and health-threatening conditions in mining, construction, manufacturing, and agriculture make for nasty, short, and brutish lives for the men as well as the women who constitute the politically and economically most vulnerable classes around the globe.

I noted earlier that wherever human interactions with nature are sexually segregated, environments are usefully conceptualized as gendered. The issue here is not that environmental preoccupations with men's issues should be replaced by a preoccupation with women's issues. Rather, the point is that addressing men's problems does not automatically address the concerns of women and their dependents, who are the majority of any society. Environmentalists have long argued that the Enlightenment entrenched a faulty philosophy of nature. Nature is not a cornucopia, available to satisfy limitless desires, as in the infant's dream of his or her mother. Moreover, sciences and philosophies of nature and of science, like all other human creations, are importantly *in* nature, not autonomous from it. Sciences, their philosophies, and their political, religious, economic, and other kinds of relations with the societies that use them should all be explained together. Yet modern philosophies' attempted isolation and immunization of natural sciences from social explanation, and their devaluation of local knowledge, have worked against such comprehensive understandings. The language about nature in scientific philosophies proposes that its principles exist outside of all human cultures and that its unique order can be identified and explained only by a context-free universally valid science. This language obscures what happens in humans' interactions with their natural and social environments. Yet such environments are by definition the only parts of nature with which humans interact, whether this nature is located en route to Mars, out past Jupiter, in the factory, in the garden, or in the kitchen. We need philosophies of environments and of human interactions with them, through laboratory as well as everyday practices, to replace Enlightenment philosophies of nature and science.

In drawing together resources from feminisms North and South, postcolonial science studies, and Northern environmental studies, Southern feminist science studies shows the importance of intellectual and political coalitions between analyses that often have been at odds with each other. Transforming Enlightenment philosophies of science requires insights from all the centers and peripheries where such beliefs have come to structure social relations and their meanings.

Conclusion

Here I have been arguing that the historical and geographical maps within which Northern science and science studies projects are conceptualized, and their assumptions about how scientific knowledge is produced, are inadequate insofar as they are still contained by Eurocentric assumptions. This is as true of Northern feminist work as it is of "prefeminist" work. Postcolonial science and technology studies have deeply challenged the exceptionalist and triumphalist assumptions of Northern sciences and their philosophies, including the field of post-Kuhnian science studies. Yet neither

Northern feminist nor mainstream sciences or science studies more generally use conceptual frameworks and practices that permit them to engage with this work. Southern feminist approaches have helped to create the resources needed by Northern feminist science and technology studies. They point to a "world of sciences" rather than to the North's exceptionalist and triumphalist preoccupation with its own supposedly one true story of nature's order.

Now we can note one more controversial feature of feminist science studies, North and South: many of their central concerns and analyses are not recognizable as scientific or philosophic ones at all from the perspective of the mainstream philosophy of science and science studies. All too often these feminist accounts seem focused "outside the realm of the true," that is, off the map where intelligible scientific and philosophic problems and research occur. Mainstream philosophers frequently dismiss them as "merely sociology," or as only about applied sciences or about technology, or only about "women's issues." Yet the feminist work cannot be shrunk into such categories. This feminist work, North and South, produces both theory about nature and society and practical, technical insights, new philosophic standards, and innovative sociological analyses, and in other ways it bridges conventional distinctions in mainstream science studies. We need to remember that what counts as nature and as the social, as well as the definition of real science, has continually changed throughout history. Science is dynamic.

Finally, there are surprising affinities and also startling political dissonances between much feminist work and the new processes of producing scientific knowledge. The "nature of science" seems to be entering a major transformation yet again in ways that bear a complex and troubled relation to feminist science studies, as well as to the science projects of other social justice movements. Is this feminist work North and South merely making contributions to the cultural logic of late capitalism, in Fredric Jameson's phrase?[17] Or are these feminist projects, along with those of the other social justice movements, slowly and painfully reempowering forms of social accountability that escape the market and can advance social justice?

PART 2

Truth, Relativism, and
Science's Political Unconscious

7

The Political Unconscious
of Western Science

Almost five decades of postpositivist science and technology studies in the North, including feminist accounts, have permanently undermined central assumptions of conventional philosophies of science. One of their most important findings is that it is impossible in principle, not just difficult in practice, to produce "pure sciences," completely devoid of any social and cultural values, interests, or other such features. This realization has enabled us to think abut the kinds of economic, social, cultural, and political assumptions and ideals that have helped to shape the sorts of sciences originating in the Global North. We can explore the political unconscious of modern Northern sciences.[1]

Such political assumptions and ideals—political philosophies—are not mere surface blemishes or decorations on sciences that are otherwise value neutral. Rather, they help to determine the kinds of scientific questions that will be asked, the ways good scientific method is conceptualized and practiced, and for whom the results of research will be especially useful. Philosophies of science shape and are shaped by social relations, though usually "behind the backs" of the philosophers and scientists who take themselves to be conducting value-free research and argument. If we can articulate such political philosophies, then we have the possibility of critically examining them, which is what the sciences have argued is the only rational procedure when weighing the adequacy of beliefs.

The argument here will be that the mainstream philosophy of modern Northern sciences exhibits an ambivalent political unconscious. It has both features that are prodemocratic and committed to social justice and others that are antidemocratic and authoritarian. Moreover, even the prodemocratic features are ones that have come under sustained criticism in recent political philosophy because they are not democratic enough. They belong to liberal democratic theory, which has proved far too useful to capitalist, androcentric, racist, colonial, and imperial economic strategies.[2]

This situation raises a number of questions. Such a philosophy of science is epistemologically and metaphysically underdeveloped in that it lacks the kinds of critical social science resources necessary for scientists and others to detect and evaluate features of the world, sciences, and their philosophies that shape the results of research. Should scientists and philosophers of science be trained into political philosophy to enable them to recognize and critically examine the political unconscious of various scientific projects? Should political philosophers learn how to critically examine sciences and their philosophies, too, among the sites where decisions with political outcomes are made? What kind of social justice theory should modern sciences and their philosophies promote? We took up such questions in several ways in earlier chapters. Here we address them from another angle.

Issues of External versus Internal Democracy

Most people concerned with strengthening the links between modern sciences and democratic projects have focused on issues conventionally conceptualized as external to sciences' cognitive, technical core. How are the economic benefits and costs of the production of scientific information distributed within societies and between them? Who receives the social and political benefits, and who must bear the costs? Who gets to make the decisions that produce such distributions? Are the processes responsible for such distributions democratic?

According to this external democracy view, sciences' internal core—their most tested theories, models, methods, descriptions, and explanations of nature's order—cannot be implicated in such questions or their answers. These scientists and critics have been concerned, for example, with who has access to mathematical, technological, and science training; who gets to decide which scientific projects should be funded; who gets access to the information and technologies that research makes possible; and who gets to make decisions about the social and environmental risks generated by scientific and technological projects. We saw such issues raised in earlier chapters. External issues about whether sciences are used in democratic ways are assumed to have nothing to do with the content of scientific claims or with its standards for good method, objectivity, and good science.

Such apparently only external issues arise in both global and national contexts. They have emerged in controversies over the resources that sciences' provide (intended or not by scientists) to militarism, to increasing environmental destruction, and to the negative consequences for the world's economically and politically least-advantaged groups of development policies and practices. For example, U.S. funding for the natural sciences—physics especially—has been disproportionately tied to national security priorities. Indeed, from the beginnings of modern sciences in Europe military interests have been the single greatest spur to the advancement of scientific and technological knowledge. The computer and frozen foods are just two such contributions of military research to everyday life. Yet the military information and technologies produced today and perhaps in much of the past end up disproportionately used within (or against) Third World societies in Latin America, the Middle East, and Africa.

In another case, post–World War II development projects were supposed to enable Third World societies to reach the higher standards of living available in the First World. Development was to be accomplished through the transfer to the Third World of First World sciences, technologies, and their philosophies of rational inquiry and rational organization. Yet the priorities, policies, and practices of these programs have ended up "developing" primarily the already most economically and politically well-positioned groups in the Global North and Global South. They have largely dedeveloped and maldeveloped the great majority of the world's peoples who are already the most economically and politically vulnerable. In most cases development policies and practices redirected the South's natural resources and human labor to serve the needs of transnational corporations and "the investing classes" globally, and substituted socially and environmentally destructive lifestyles for less harmful local practices.[3] (We looked at this situation in earlier chapters.) Modern sciences' agendas have often ended up aligned with antidemocratic projects globally as well as nationally, though this certainly was not the intent of most of the scientists or, in many cases—perhaps most of them—of development administrators. For the most part, preserving or advancing (or both) desirable cultural, political, and environmental values was simply not on the agenda of agencies or their funders who equated development with economic growth.

These external approaches, crucially important as they are, appear not to challenge the idea that social and political neutrality can, does, and should characterize sciences' internal cognitive, technical cores. They appear not to challenge the Enlightenment assumption that sciences can be, should be, and in the best existing cases are value free. Of course, modern sciences are conducted within social worlds in that their human and material resources must be provided by the larger social order, as defenders of this view readily admit. Moreover, the amounts, kinds, and sources of such resources have varied from era to era and from one culture to another. But sciences' transcultural, socially neutral theories, models, and methods enable them to detect the facts about the order of the universe that are everywhere and always the same. According to the externalists, sciences are in society, but society is not *in* sciences and their best theories, models, methods, or results of research.

This is odd, since many historians of modern Western sciences have consistently pointed to distinctive ideals of democracy embedded in scientific principles and practices. Some have noted authoritarian ideals there too, perhaps a residue of modern sciences' medieval clerical culture.[4] Today, cognitive democracy approaches, as I will call them, are concerned with how social and political fears and desires get encoded in that purportedly purely technical, cognitive core of scientific projects. They are concerned with how society gets inside sciences. They want to have something to say about just which aspects of society get so encoded, becoming powerful ideals that then in turn have political effects upon their social environments. These approaches ask if the external features of sciences' environment really have so little influence on its internal processes and their outcomes—on the content of sciences' claims—and what the external effects are of internally coded political philosophies. They explore the "unconscious" of modern sciences and their philosophies. This issue has been raised in other ways in earlier chapters. This chapter tries to identify both pro- and antidemocratic internal values in modern sciences' cognitive, technical core. What is the best way to engage with this kind of phenomenon in trying to link sciences more closely to democratic projects?

However, before turning to say more about it, three possible misunderstandings of this kind of project must be addressed. Some readers may fear that this cognitive approach adopts a relativist epistemology, or consists in a "flight from reason."[5] There are no sound reasons for such fears. The cognitivists' point—at least as I and most of such analysts develop it—is not that scientific principles and practices are "nothing but" social, political projects or that the representations of nature that they produce are shaped entirely and only by such projects. Rather, the point is that technical, cognitive elements of scientific practices and the information they produce always represent social and political priorities, meanings, and ideals as well as more or less accurate pictures of nature's order. There are indeed rational theoretical and practical standards for evaluating competing knowledge claims. Modern sciences do in many obvious ways achieve fewer and fewer false claims about nature's order. But that does not mean that they are uninfluenced by social phenomena.[6]

Second, concern with philosophies of science may seem to many readers largely irrelevant to the real-life projects of encouraging closer links between modern sciences and democratic policies and practices. My point is that it is precisely such encoded ideals that have powerful effects on our daily activities. Such ideals make seem natural, logical, "commonsense," and otherwise desirable precisely the kinds of antidemocratic policies and practices of concern to the externalists. There are consistencies between the external antidemocratic policies and practices that have shaped modern sciences and models or idealizations of such policies and practices that can be found in philosophic aspects of sciences' cognitive cores. Moreover, this argument makes another point: there are significant scientific costs to such antidemocratic idealizations as well as the more obviously recognizable political costs.

Another caveat. Perhaps such concerns will appear to *introduce* social and political elements into otherwise socially neutral sciences and accounts of them. As will be clear from what follows, however, everyone who reflects on the matter understands that modern sciences already do encode precisely such social and political fears and desires. The question here, instead, is whether they should more effectively encode democratic ideals, how they can do so, and on what grounds one could justify such recommendations. A preoccupation with the futile (and undesirable, the argument will go) project of excluding social and political fears and desires from the sciences' cognitive cores delegitimates and distracts from these kinds of important issues.

There are many respects in which philosophies of science could and do encode both pro- and antidemocratic ideals. Here I shall briefly focus on just one philosophic notion that brings together a number of antidemocratic elements, enabling us to highlight contrasting prodemocratic ones. This is the unity of science thesis and its universal science ideal. But what do or should I mean by democratic ideals, by "democracy"? Section 2 distinguishes three meanings of the term that are relevant to a discussion of sciences. Section 3 summarizes the unity of science thesis and its scientific and political costs.[7] The following two sections identify some anti- and then prodemocratic ideals encoded in conventional philosophies of science. The concluding section proposes another way to think about the value of universalizing scientific hypotheses that can strengthen prodemocratic ideals and projects.

Democratic Ideals for Sciences

Skepticism about the ideal of democracy is widespread these days, as noted above, especially when such an ideal is advanced by someone of European descent. The term has been captured by conservative U.S. and European politicians who insist on restricting the democratic practices they would install at home and around the globe to those practices that make the world safe for expanding capitalist projects. But the term has too valuable a legacy to be abandoned to such antidemocratic politicians. Moreover, even the tiny democratic opportunities these politicians would permit create opportunities and desires that are not easy for them to control. Politically and economically vulnerable peoples around the globe yearn for fully democratic social relations. What political standards should one use for selecting democratic social ideals for the sciences?

One approach would be to identify a general democratic principle that is relatively uncontroversial among prodemocratic theorists everywhere, and then try to specify social practices appropriate to local conditions of scientific research that would conform to it. One candidate for such a principle could be the familiar claim that those who bear the consequences of decisions should have proportionate shares in making them. This would be one way to guide the selection of scientific practices that are responsible, accountable, to those they affect. There will be exceptions made to such a general rule, of course: children and other very carefully identified groups cannot be expected to be able to make such decisions, or to make them wisely. Specification of the exceptions will itself be an important part of any democratic process. Nevertheless, the general principle is attractive because it has guided so many different kinds of effectively democratic practices. The institutions and procedures through which those proportionate shares would be exercised can be expected to vary in different social contexts: practices appropriate for small, homogeneous, or oral cultures would have to be different from those appropriate for large, heterogeneous, and multiply-literate cultures.

Political philosophers point to the varying democratic effects of three kinds of more specific principles that have been thought to conform to such a democratic directive. One recommends that the interests of relevant groups should be fairly represented during decision-making processes. In local, national, and transnational scientific councils, the interests of all the groups who will bear the consequences of scientific and technological decisions should be represented during policy-making processes. Although this conception of democratic practices is better than none, critics argue that it does not produce democratic-enough effects. Who is to represent such interests? Should those benefiting from their exercises of power be presumed to be able to represent fairly the interests of those over whom they exercise such power? And how are such councils to be held accountable for identifying who the relevant groups are, how their interests can best be fairly represented, and how democratically to resolve conflicts between competing interests? This principle has been used to justify patently unjust social and political systems; it has often been accused of a paternalism that blocks recognition and appropriate consideration of the interests of the politically and socially least-powerful groups.

A stronger proposal is that members of such relevant interest groups should themselves have rights to represent their groups' interests in decision-making councils: there should be proportionate numbers of women and men, whites and blacks, and others who will be affected by a particular scientific project among the groups that design, fund, and manage such projects. This practice can go far to correct antidemocratic tendencies in any project. There is ample evidence that not only scientific and technological benefits but also political ones can develop from grassroots organization, "participatory action research," "bottom up" design, and other such ways of giving "end users" a central voice in the design of scientific and technological projects. Nevertheless, these approaches can be enacted in too-conservative ways that reduce both their scientific and their political benefits. Widespread experience with "adding women and minorities" to work sites and policy groups where they have heretofore been excluded reveals such limitations. Who decides which groups have a right to be so represented? Will only the least-threatening members of minority groups be the ones permitted into science policy councils? Must they suppress their "difference"—mute their demands or present only those parts of them that easily fit into prevailing concepts and practices—in order to function effectively within the prevailing standards for organizational behavior in scientific and technology institutions? Can individuals produce powerful-enough discourses on behalf of their groups' interests to compete effectively with the prevailing dominant discourses, institutions, and practices of the majorities?[8] After all, it took decades of political and intellectual work by many diverse groups to produce evidence and arguments compelling to international agency administrators demonstrating that so-called development was primarily delivering benefits to already advantaged groups and dedeveloping and maldeveloping the already least advantaged. This understanding was not one that automatically became visible even to those initially disadvantaged by development policies.

The strongest proposal for achieving democratic standards is that these require real equality among groups in the institutions and societies in which decision making occurs. Until sexism, racism, and class systems no longer are able to distribute social, political, and economic resources inequitably, institutions within such societies cannot achieve maximally democratic decision processes. "Real equality," moreover, includes symbolic as well as material resources. Thus, if a culture's ethnocentric standards—the inequality of groups, their thinking, and traditions—are modeled as ideal in the cognitive cores of sciences, one should not expect real equality among members of such groups in scientific institutions and cultures. As long as objectivity, rationality, and the practice of good method are defined against whatever is associated with the feminine or "primitive," the perspectives of women and the modern West's Others cannot gain an equal footing in scientific decisions. Insofar as the politics of the larger society are modeled in a science's cognitive core, they serve as obstacles to efforts to eliminate other ways in which such standards shape scientific practices. Attempts at explicitly more democratic decision processes in scientific institutions are frustrated by models of ideal scientific practices that "unconsciously" broadcast antidemocratic messages. The latter are all the more powerful when they are obscured in a cognitive, technical core of science claimed to be immunized against the possibility of social influence and, thus, of the need for critical social analysis.

Thus, more science and technology in undemocratic societies cannot be expected to deliver the benefits and costs of research democratically, according to this view. In such conditions, more science and technology are guaranteed to increase social inequality, and this is so in spite of the intentions of individual scientists or policy makers. Scientific institutions and their cultures and practices cannot by themselves counter antidemocratic power distributions within society's other institutions that ensure that only those already politically, economically, and socially advantaged will be in positions to be able to take advantage of the information that scientific and technological work produces. When scientific institutions, cultures, and practices produce conflicting political messages, it is even more unreasonable to expect them to be advancing democratic social relations. There are scientific costs to such antidemocratic conditions also, to which I turn in sections below. This situation predicts a rather depressing scenario to most of us who thought that more science and technology always delivered at least some benefits "to humanity," however much other benefits were siphoned off by the already overadvantaged.

Yet we cannot wait for "real equality" to conduct the kinds of scientific research that social justice movements need. Women, the poor, and exploited racial and ethnic groups around the globe need more reliable information about the threats to health and life that they face daily.[9] The unity of science thesis is one important site where, it turns out, various antidemocratic ideals come together, with effects on sciences' abilities to represent and serve human interests democratically, to include oppressed groups in scientific processes, and to advance full social equality.

The Unity of Science Thesis: One World, One Truth, One Science?

Elsewhere I have discussed this familiar model of nature and science.[10] Here I briefly summarize this thesis and then turn to its "unconscious" antidemocratic message.

This model of the sciences and the world they would describe and explain became popular in the late nineteenth and early twentieth centuries; it became a kind of intellectual movement in the first half of the last century. In modified versions it still holds great appeal for many philosophers, scientists, and the general public. However, today it is also the object of considerable skepticism on the part of philosophers and historians of science.[11] Moreover, it is challenged also by feminist and postcolonial insistence that sciences are always socially situated and that the world contains many such socially situated sciences, as we saw in earlier chapters.

According to the unity argument, there is one world, one and only one possible true account of it, and one unique science that can capture that one truth most accurately reflecting nature's own order. Less visible in most articulations of the unity thesis is a fourth assumption: there is just one group of humans, one cultural model of the ideal human, to whom nature's true order could become evident. For early modern scientists and philosophers, the ideal human knower could be found among members of the new educated classes. Such individuals could use distinctive knowledge-seeking procedures, and thereby their theories could come to reflect the true order of nature that God's mind had created, just as the latter had also created human minds "in his own image." As the ideal human mind came to occupy the place in modern philosophy that the soul had occupied in Christian thought, rational man replaced spiritual man as the chosen recipient of the one true vision of the world's order.

Today, the influential Harvard biologist Edward O. Wilson has mounted a renewed appeal in *Consilience: The Unity of Knowledge* for what he refers to as "the Ionian Enchantment." This is "a conviction, far deeper than a mere working proposition, that the world is orderly and can be explained by a small number of natural laws." He continues: "The central idea of the consilience world view is that all tangible phenomena, from the birth of stars to the workings of social institutions, are based on material processes that are ultimately reducible, however long and tortuous the sequences, to the laws of physics."[12] Whatever philosophers, historians, sociologists of science, feminists, and postcolonial critics may think about such a "worldview," it clearly is compelling to many scientists, in popular science debates, and apparently in science policy circles. Perhaps it is precisely the deep unease that these new "sciences of science" create in their challenge to the unique authority of the sciences of the North, and the specter of relativism that they raise for many observers, that impels Wilson and those sympathetic to his views to reassert the conventional belief that Western science provides the one and only authoritative account of reality.[13]

Recently, several studies of the thesis and its development in the twentieth century have appeared.[14] They tell a much more complex history of the thesis than writings such as Wilson's assume. Even within science studies different traditions can be seen renegotiating the terms of the debates over the unity thesis. Yet they share the conviction that there is no compelling evidence available for the form of the thesis that appears in writings such as Wilson's. For one thing, modern science in the North is plural. There are many distinctive modern sciences with incompatible ontologies, methods, and models of nature and of the research process. If *unity* means singularity rather than simply harmony, there seems to be no encouraging evidence for the possibility of reducing them to one methodologically, ontologically, theoretically, linguistically, or in any other meaningful sense of that project.[15] Perhaps *unity* should be taken to mean only harmony, as some early defenders of the unity thesis had in mind. However, although many kinds of such harmony certainly do exist among the sciences, such an interpretation of unity undercuts attempts to claim universality for elements of the cognitive cores of sciences. There can be all kinds of "harmonies"—sharings, borrowings, echoes, communications—of disparate elements without any elements at all achieving universality. The notion of harmonious sciences is too weak to do the work of the unity of science thesis.

Yet the goal of *universalizing* one's hypothesis—of trying to grasp the limits of the domain of its usefulness—remains important in any scientific process. Mathematical expression has been one important such universalizer. Can the universalizing ideal be detached from the attempt to discover or create just one legitimate scientific tradition, that is, from the unity of science ideal? Let us hold this question until we have reviewed the scientific and political dysfunction of the unity of science ideal itself.[16]

Costs of the Unity Ideal

Arguments in the preceding sections and chapters have already pointed to some of the most important political costs of the unity ideal. First, it supports the devaluation of forms of knowledge seeking that have proved valuable in other cultures, indeed of ones that today are crucial to the survival of groups effectively delinked from the benefits of international science and technology. Some African, Asian, and South American cultures, for example, have little or no access to international science and technology, yet they survive—and in some cases thrive—thanks to the strengths of their local knowledge traditions.

Second, to devalue these traditions is to devalue the people and cultures that use them. This legitimates the continuing forcible subjugation of these groups to Western projects—military or commercial.

Third, the unity ideal supports the construction of models of the rational, the objective, the progressive, the civilized, and the admirably human in terms of distance from the non-European, the economically frugal, as well as the feminine. The authority of the unity ideal is presented as a necessity for the distinctively rational, progressive, civilized, and human.

Fourth, it insists that science speak in a monologue, thereby elevating authoritarianism to a social ideal, when it asserts that it is desirable for everyone to acknowledge the legitimacy of one culture's (the "international science culture's") claim to provide the one true account of the world.

This "political unconscious" exacts scientific costs. First, the universality ideal legitimates decreasing cognitive diversity, yet it is just such diversity that has provided continuing resources for the growth of every culture's scientific and technological projects. Cultures need other systems of knowledge from which to borrow novel understandings of local environments and the resources they can offer, as well as metaphors, models, narratives of nature and humans' place in nature, new inquiry techniques, and ways of organizing the production of knowledge. Otherwise, any knowledge system would be stuck with only what can be generated from within its own culture. Modern sciences would have been deeply impoverished without the resources gathered into them from the knowledge traditions of the other cultures encountered by Europeans. Moreover, we cannot know what knowledge we will need in the future as social and natural environments change and new needs and desires develop. Different cultures' constantly evolving ways of thinking about nature and social relations will continue to provide valuable resources for each other's projects. It is as foolish to decrease cognitive diversity as it is to decrease biological diversity.[17]

Second, the universality ideal legitimates accepting less well-supported claims over potentially stronger ones in many cases. If the aspects of the world on which a claim focuses, the methods used to gather evidence for it, or the models and narratives through which it examines nature do not fit with the one prevailing one, it can be ranked as less probable than a claim with far weaker empirical evidence but which is consistent with prevailing scientific models. Thus, acupuncture's ability to manage chronic pain was dismissed as mere superstition or folk belief regardless of testimony from grateful patients until neurochemical analyses finally identified the aspects of this procedure that could be understood in the kinds of terms favored by Western biomedicine. Environmental studies that rely exclusively on the analyses of physical sciences cannot recognize as valuable components of "the best explanation" the kinds of analyses that social scientists bring to environmental studies.[18] And, of course, feminist and indigenous knowledge claims, regardless of the quality and quantity of the empirical evidence for them, are dismissed as dangerous nonsense by many scientists.[19] Of course, it is valuable for each culture to test the claims of others within the resources of its own knowledge system. What is problematic is to assume that such a procedure correctly identifies the absolute worth of a claim, rather than only its ability to be confirmed or disconfirmed within a favored knowledge system. Here is where the threat of a disabling relativism rears its head for many scientists. (This issue is addressed below and in chapter 9.)

Thus, in the third place, the unity ideal legitimates resistance to some of the deepest and most telling criticisms of particular scientific claims. Criticisms that cannot be recognized as coming from within established boundaries of scientific discussion are perceived as unintelligible and can legitimately be devalued or ignored. It is not that they appear false, which they do; worse, they seem "outside the realm of the true." Thus, feminist analyses are persistently conceptualized as irrational and as coming from outside science, even when the critics are respected scientists, as has been the case, for example, with feminist biologists' criticisms of standard interpretations of evolutionary theory and medical representations of women's bodily processes.[20] Similarly, postcolonial criticisms of Western scientific and technical expertise is often rejected as coming from outside science even when the critics are trained in Western science. This is so even when the critics' goal is not to reject Western scientific expertise wholesale, but rather to better integrate it with insights from local knowledge systems. The ability to detect "rigorous refutations" weakens when rigor is presumed to be the monopoly of the one and only real science.

Next, the unity ideal promotes only narrow conceptions of both nature and science. As long as physics is presumed to be the model for all sciences, whether on historical, ontological (for example, its focus on primary versus secondary qualities), methodological, or other grounds, other ways of understanding nature's order will be devalued. For instance, the unity ideal blocks our ability to bring into focus the social elements—institutions, practices, meanings—in what are often presented as merely natural, scientific, and technological changes. As ethnographer and philosopher of science Bruno Latour puts the point, our supposedly modern sciences conceptualize our knowledge of nature as separate from matters of our interests, of justice, and of power, though it is in fact inseparable.

> On page six [of my daily newspaper], I learn that the Paris AIDS virus contaminated the culture medium in Professor Gallo's laboratory; that Mr. Chirac and Mr. Reagan had, however, solemnly sworn not to go back over the history of that discovery; that the chemical industry is not moving fast enough to market medications which militant patient organizations are vocally demanding; that the epidemic is spreading in sub-Saharan Africa. . . . [H]eads of state, chemists, biologists, desperate patients and industrialists find themselves caught up in a single uncertain story mixing biology and society.[21]

Where are the sciences for this real world with which we daily interact? Where are the sciences of social and natural networks and hybrids that make up reality? Latour's answer is that they are to be found in the new critical science studies. These "sciences of science" have produced precisely the research that Edward O. Wilson alternately ignores and overtly dismisses.

Another limitation is that the ideal of one true science obscures the fact that any system of knowledge will generate systematic patterns of ignorance as well as of knowledge. Every knowledge system has its limits, since its priorities select which aspects of nature's order to study; which questions to ask; which metaphors, models, narratives, and other discursive resources to use; and which ways to organize the production of knowledge. Knowledge systems are like Thomas Kuhn's paradigms in this respect. They can prove illuminating far from their original sites of production, but they all have their limits and produce recognizably diminishing returns sooner or later.

Finally, such a model for the natural sciences promotes similar problems in the social sciences that model themselves on the natural sciences, such as physicalist psychologies; rational-choice theories in economics, political science, and international relations; and positivist sociologies. Moreover, the prevalence of such models in the social sciences has bad effects in another respect: social scientists who oppose such models often can see no alternative but to focus entirely on the micro and the local. Relativist epistemological positions start to look far too attractive as long as the unity ideal is the only alternative. As a reaction to naturalistic social sciences' devaluations of the local, these other social research projects get contained by the local. The unity and relativist positions are really two sides to the same coin. In effect, the unity ideal's conceptual world is advanced in unarticulated forms through the relativist positions—a topic pursued further in chapter 9.

Thus, adherence to the unity ideal brings costly political and scientific consequences. Let us step back now and consider the anti- and prodemocratic ideals reinforced by the unity of science thesis.

Antidemocratic and Democratic Ideals

ANTIDEMOCRATIC IDEALS

To start, here modern Western science speaks in a monologue. There is one and only one science that can accurately reflect the one and only way that the singular world is organized. The unity ideal values only coherence in scientific accounts, undervaluing the benefits of conflict and dissonance so visible in the history of science. In this scheme there is no room for permanently dissenting voices, for other ways of seeing and conceptualizing nature's order, unless they are imagined as only temporarily conflicting accounts that can eventually be incorporated into the one true account. Different disciplines and other cultures are not envisioned as productive of distinctively valuable sciences that could contribute to human knowledge and welfare yet retain their separate identity. If there are two conflicting accounts of any phenomenon, at least one must be wrong.

Relatedly, this voice of expertise is authoritarian. It asserts an account of nature's order that all reasonable, civilized people should recognize and honor as legitimate. Its evidential base of observation and reason is to be regarded as unchallengeable. Indeed, the ability of an individual or culture to recognize the authority of this voice of expertise has persistently been taken as in itself a mark of rationality and civilization.[22]

The legitimacy of this authoritarian monologue is protected by the insistence of natural sciences and their philosophies on excluding resources for critically analyzing the social and political relations that have shaped the goals, processes, and effects of Western modern sciences and technologies. As long as the cognitive, technical core of natural sciences could be assumed to constitute the only elements of scientific processes that can count as "real science," and these were presumed to be free of any and all cultural fingerprints, such an exclusion did not matter. But once it became impossible to support these two assumptions either empirically or theoretically, the need for critical social resources became urgent. The most obvious sources of such tools are the critical social sciences, which natural scientists, many philosophers, and even conventional social scientists take to be "outside science." The gaze of the critical social sciences on the natural sciences and their philosophies is regarded as a threat to the purity of the latter; in such activities the social sciences are accused of introducing politics into sciences that had been heretofore free of them. Criticism from within legitimate scientific communities is always welcome, according to the dominant philosophies of science. It is crucial for the growth of knowledge. Yet the borders of such communities must be tightly patrolled against criticism from outside the scientific community. Ideas and criticisms from individuals or groups outside these communities need not be given serious consideration. Targets of such devaluation in the past half century have included Rachel Carson[23] and subsequent environmentalists, Marxists, feminists, and any researchers whose work is conducted within cultural frameworks other than those preferred by now-global elites. The continued preoccupation in defending the border between science and "pseudoscience" keeps a barracks mentality in place inside Western modern scientific and technology institutions and their philosophies.[24] As Gibbons et al. put the point, this stance leaves the natural sciences and their philosophies epistemologically underdeveloped.[25]

The persistence of these antidemocratic elements in Western modern sciences' cognitive, technical core seems to be an example of what Thomas Kuhn identified as normal science models of good science. Recollect Kuhn's argument that it is a sign of their maturity when sciences stop listening to every kind of criticism their work raises, and instead focus on working out the details and how to expand the domain of promising new kinds of research. The kinds of communities of scientists Western modern sciences have created are ones trained to make the best judgments about what is a relevant criticism of their work and what is not. Recollect also that Kuhn's critics were quick to point out that these were antidemocratic ideals he was praising and that his justification of them was not as compelling as he imagined.[26]

These antidemocratic ideals are bad enough. But let us look at the way historically prodemocratic ideals are weakened or restricted in the unity thesis.

PRODEMOCRATIC IDEALS

THE EQUALITY OF OBSERVERS   For the early modern world in which Western modern sciences emerged, the ideal of epistemic equality of sciences' observers, be they aristocrats or artisans in their everyday lives, was a radical break from the practices of the premodern world. This ideal is part of scientific methods: the social status of scientists is an irrelevant standard to invoke in judging the adequacy of scientific work. Modern science can flourish only in democratic communities, the argument goes, for only such communities can foster the critical attitude toward accepted belief that the advancement of scientific knowledge requires. Galileo argued that through his telescope anyone could see the facts about the heavens. Today, the sciences still cultivate practices that support this equality ideal. For example, a graduate student, no less than a Nobel Prize winner, is encouraged to offer observations and arguments.[27] Of course, experienced observers can be expected to come up with more reliable contributions. However, the persistent emergence of valuable observations and arguments from less or, better, differently experienced observers sustains belief in the scientific value of this democratic ideal.

Yet in societies with substantial economic, social, and political stratification, there can be only a formal equality of observers. Women, the poor, devalued minorities, and the majority of the world's cultures have been excluded from opportunities to engage in what the West counts as science and, more important, to direct national and global science and technology policy. And even when they are actually included, it has been easy to keep their voices silenced or, at least, ineffective. Today, protests against this exclusion and the economic conditions ensuring it have become globally visible—most recently in feminist organizing and in uprisings inside and outside the World Trade Organization and World Bank meetings. Feminists have insisted that women's voices count in decisions about biological and medical research, in environmental studies, and especially in the debates over sustainable Third World development. At the above-mentioned protests, poor countries are insisting that their voices count in global trade and development policy, not just the voices of the rich countries. Third World development was from its origins defined as the transfer of Northern sciences and technologies and their rationality to the "underdeveloped" societies of the South, so Northern sciences and technologies are deeply implicated in these protests against global trade and development policy.[28]

However, there are additional problems with this ideal. Its emphasis on what "anyone can see" assumes that reliable observers are always only individuals and that scientific facts are entirely out there in socially unmediated nature, outside social and cultural frameworks and practices. These assumptions work against recognition of the scientific value of cultural differences in what people do see. If the prevailing standards of Western elites ensure that they do not see the realities visible to other groups, others' claims do not get to count as rigorously produced knowledge.

Yet, as was discussed in an earlier chapter, it is now clear culture can be productive of knowledge, not just an obstruction to it—which, to be sure, is often the case. Different cultures have different locations in nature's heterogeneous order, and can have different interests even when occupying "the same" environmental niche. They bring different discursive resources—metaphors, models, narratives—to interactions with their natural and social environments, and they tend to organize the production of knowledge in ways typical of how they organize other forms of labor. Finally, a culture's location in political relations, local, national, and global, both enables and limits the kinds of resources available to it of the four preceding kinds: poor people and rich people, men and women, and slaves and their masters can each tend to produce distinctive understandings of nature and social relations. The point here is not that the reasons for understanding equality as exchangeability are without appeal. The idea was that if differences in nature's order are to be identified and confirmed by successive observations, then social elements of observation must be held constant; they must be eliminated from having effects. Rather, the issue is that even though this way of thinking about the equality of observers is appealing, it is not effective at excluding all cultural values and interests from the results of research, and it excludes recognition of the value to the growth of knowledge of cultural cognitive diversity.[29]

This leads us to the situation that the unity ideal encourages the elevation to universal status of what are distinctively Western sciences. Western sciences can achieve unique universality only in an authoritarian way. As critics have pointed out, the appearance of culturally neutral sciences is precisely what best serves the interests of the transnational corporations and agencies that now are major features of the landscape within which Western modern sciences function. These institutions need sciences and technologies that are free enough of local cultural features to appear culturally uncontroversial around the globe, yet have the distinctive cultural features useful for organizing and propelling the current capitalist, antidemocratic form of the global political economy.[30]

UNIVERSALLY HOMOGENOUS MATTER   To turn to another democratic feature of science that also has an antidemocratic side, in early modern science the belief that matter was everywhere composed of the same kinds of materials was perceived to be a radically democratic claim. It challenged the Christian argument that the celestial and terrestrial spheres were composed of fundamentally different kinds of matter that obeyed different kinds of laws. To challenge the hierarchy of matter was to challenge also the political hierarchies that the Christian view of matter modeled. Christian hierarchies began to lose their ability to model the "order of nature," and vice versa, after the advent of early modern science, for the matter composing the earth and the heavens increasingly was seen as equal—indeed as identical.

This ideal has led to the practice of treating as real only what Western modern sciences can observe, or, in actual practice, are interested in observing. Since Western modern sciences make a point of excluding rather than systematically observing aesthetic, moral, spiritual, social, cultural, and political phenomena, these are assigned a lesser reality, if any, than the objects of scientific scrutiny. Other cultures' ways of intellectually engaging with such phenomena are regarded as lacking scientific rigor, as all too permeated by the social and cultural assumptions the natural sciences have been concerned to eliminate from their own analyses. As forms of scientific rationality permeate a culture, all of nature and social life gets drawn into the logic of scientific accounting. Moreover, the critical skills necessary to analyze such a social process cannot gain legitimacy. Thus, concerns to rationalize social processes need not be troubled with the positive effects of aesthetic, moral, spiritual, social, cultural, and political conditions and meanings on the growth of knowledge, or with their value in themselves. Here, too, what was once a powerful prodemocratic feature of modern Western sciences has turned into one supporting "might makes right."

PUBLIC ACCESS TO SCIENTIFIC PROCESSES AND RESULTS   Another important prodemocratic feature of scientific rhetoric is commitment to the public status of both scientific methods and results of research. The results of research must be public; they belong to "humanity" and on scientific grounds may not legitimately be shielded from public view. This requirement has been a focus of the external analyses. However, it also plays a crucial role in scientific method. This ideal permits other scientists to repeat experiments and observations so as to identify results of research that were peculiar to local conditions of inquiry. Thus, this replicability requirement is part of fair-inquiry procedures, publicly reported and publicly observable. The sciences, in contrast to antidemocratic institutions, are to be transparent. Their own processes are to be open to critical scrutiny. This is thought to ensure that these processes are not visible in the results of research themselves, that is, that they do not make contributions to what is "out there," observed. In the results of research we are to see only nature's regularities and underlying causal tendencies, not the technology of inquiry itself. Moreover, a general familiarity with these procedures of fair inquiry empowers citizens of Western modern societies to question the ignorance and superstitions upon which the legitimacy of antidemocratic power depends. This happens as people learn the good effects of relying on their own powers of observation and reason, which the practice and appreciation of Western modern sciences encourage. Thus, the argument goes, Western sciences help to create knowledgeable individuals who are willing to insist that social life be guided by the kinds of empirical fact-gathering and rational processes that have been so successful in Western sciences.

Of course, the "public" that in principle has access to the results of research is not in reality substantively democratically organized; it is not composed of groups with equal social power. Powerful private corporations and national and international agencies, ones that have a monopoly on the resources necessary to replicate the processes through which Western modern sciences produce information, also share distinctive cultural values, interests, and conceptions of nature and inquiry. Thus, the fact that the results of research have been replicated within these elite communities can guarantee only their transcendence of values and interests that differ within such communities.

Public access to the results of research has become severely limited through legal systems of patents and contracts. It is also limited by the emphasis on experiential and tacit knowledge, and decreased interest in publication, in some particular powerful new research processes. And it is limited by the real inequality in and between global societies that puts basic scientific literacy out of reach of the world's most economically and politically vulnerable populations.

There are additional once prodemocratic but now antidemocratic features of modern Western sciences' "political unconscious." For example, critics have pointed out that rationality, possessed by all humans, is in philosophies of modern science restricted to a purely instrumental reason. Jurgen Habermas, feminist critics, and critics from other cultural traditions have commented on the unfortunate neglect of broader and different kinds of reason that play a productive role in human knowledge seeking. Emotional, cultural, aesthetic, religious, social, bodily, and political reason have contributed to the growth of knowledge but are devalued in favor of only the instrumental expertise of scientific elites.[31]

The "political unconscious" of Western modern sciences is a full and conflicted one. What should and could be done about transforming it in social justice directions?

Rethinking Modern Western Sciences' Political Unconscious

I have been arguing that conventional philosophies of modern science contain encoded political ideals—a political unconscious. Some aspects of this politics appear antidemocratic from the start. Others have a history of prodemocratic meanings, but have been severely weakened and are restricted today. Modern Western science began as a guerrilla warrior challenging powerful religious and state interests. Today, it is deeply enmeshed with powerful global economic and state interests. As one observer has noted, science has itself become a "reason of state," invoked by states to claim authority in global circles.[32] Many observers have noted how the invocation of scientific expertise has become a substitute for democratic politics. Decisions that have huge effects on how we live (and die) are made by bypassing democratic decision processes in favor of scientific and technical decisions.[33] We should expect that the political unconscious of such a science would not be a pretty sight to those seeking to expand democratic processes.

Yet this is too gloomy an assessment. Much work has already strengthened both the scientific and the prodemocratic power of modern Western sciences' ideals. Objectivity, rationality, and "good method" have been three concepts that have received special attention.[34] Note that the goal in these writings has been not to distance science from such notions, but to increase the competence of science's political philosophies at advancing both the growth of knowledge and social justice.

Another such focus has been to imagine a substitute for the unity ideal. What would it mean to give up the ideal of one and only one perfect knowledge system and, instead, imagine a world of sciences? Such sciences would partially overlap with each other in various ways through borrowings and shared interests, yet also each retain the kind of distinctive cultural identity that has proved so fruitful for the growth of human knowledge in the past. Earlier chapters have tried to identify elements of such a vision. More work is needed, and it cannot wait until we have a substantively egalitarian world but, rather, must begin to emerge now to assist the birth of such a world.

And here is just one way that philosophies of modern sciences could encode democratic ideals more effectively with benefits both for democratic social movements and for maintaining the diverse cognitive resources every knowledge system needs to flourish.

8

Are Truth Claims in Science
Dysfunctional?

Do we need truth claims? Can they have ethical and political consequences, or do no facts imply norms or "oughts," as philosophers have
maintained? To be sure, there are contexts in which truth claims serve valuable purposes. However, the argument here is that in the case of the sciences,
their costs appear to outweigh their benefits. Whereas conventional philosophies of science and popular thought have assumed truth claims to be an uncontroversially valuable goal for the sciences, a critical evaluation of this assumption has emerged in the past four decades from three schools of science
studies: Euro-American philosophy, history, sociology, and ethnography of
sciences; feminist science studies; and postcolonial science studies. From the
perspectives of central themes in these accounts, the ideal of truth obstructs
the production of knowledge. Moreover, claims to truth support antidemocratic tendencies in science and society because a democratic social order in
a multicultural world should not provide the necessary conditions for the
kind of strong, universal agreement among scientists that the truth ideal requires. The truth ideal in science supports tendencies toward inequality.

It should be noted immediately that all of these schools of science studies hold that it is possible and desirable to avoid adopting an epistemological relativism that forecloses the possibility of rationally sorting beliefs, as
we can see in the next chapter. Giving up the truth ideal does not force one
to an epistemologically damaging relativism. Obviously, these are complex
and contentious matters. The evidence for such claims has arrived from careful studies of scientific traditions and practices, as well as from political philosophy. Here we can outline the arguments in a way that makes apparent

the necessity of critically reevaluating widespread rhetorical reliance on truth claims if the sciences are to avoid advancing inequality.

Earlier Skeptical Accounts: Quine, Duhem, Popper, Kuhn, and Feyerabend

Let us begin with arguments from European American studies of science. One immediately notes that skepticism about the truth ideal is by no means new, that it can be found, among other places, in the work of philosophers and historians regarded as mainstream and perhaps even old-fashioned. Back in the 1950s philosopher W. V. O. Quine argued that analytic and synthetic claims were not as independent of each other as had been supposed. Scientific hypotheses cannot meet the tribunal of experience, as he put the point, one by one, but only as parts of networks of largely unarticulated everyday and scientific assumptions about the reliability of experimental instruments, of particular observers' sight and hearing, of principles of optics, of assessments of prevailing environmental conditions (such as the purity of the water or air), of expectations about possibilities of divine intervention, and about many other matters—including even the laws of logic.[1] The laws of logic are supposed to be purely analytic, yet Quine is here arguing that empirical observations could, when all other attempts at revision of belief seemed to lack plausible support, justify changing them or, at least, setting them aside in a particular case. Quine's point is that scientific thought and everyday thought form a seamless web of belief. In each scientific experiment, though scientists focus on only one or a few hypotheses, in effect the entire network is tested all at once.

In his development of such insights, Quine was pursuing Pierre Duhem's arguments from early in the century against the possibility of a crucial experiment that could simultaneously prove true one hypothesis while disconfirming all others. Duhem argued that because of the role that background beliefs played in testing procedures, and the necessity of holding even well-confirmed hypotheses open to the collection of possibly disconfirming further evidence, no hypothesis could be conclusively confirmed or proved true.[2]

Meanwhile, Karl Popper argued that although hypotheses could never be confirmed, as the inductivists imagined, they could be conclusively disconfirmed. Indeed, the best strategy for scientists was to think up the most severe tests for hypotheses, ones that only the strongest hypotheses could survive. For Popper, inductivism could not explain modern science's successes, but deductivism—"conjectures and refutations," as the title of one of his books put the matter—could. He also argued that it was a sign of a dogmatic attitude to think that one's beliefs were true and thus irrefutable; religion, Marxism, and psychoanalysis were his examples of belief systems lacking the possibility of empirical falsification. Scientific beliefs contrasted with such dogmatically held beliefs in that scientists always insisted on holding beliefs open to the risks of possible future falsification.[3]

Quine went further, however, to argue that useful as severe tests might be for advancing the growth of knowledge, deduction could never produce absolute disconfirmations of any single hypothesis by itself. Nature's "nay," as he put the point, always left scientists free to attribute the apparent mismatch between scientific statement and nature's order to any of innumerable background assumptions or even to the empirical observations no less than to the articulated hypothesis on which attention had been focused. Some background beliefs would always appear more reasonably doubtful than others. However, when all such reasonable doubts failed to turn up a culprit, yet additional background beliefs, initially beyond suspicion, would be examined for the possibility of revision. To put the point another way, empirical observations, theoretical hypotheses, and background beliefs could never gain the independence from each other that the conventional "logics of justification" required. Even the laws of logic have final justification in their usefulness for helping us make sense of the world around us.[4] However, although neither "true" nor "false" could in principle justifiably be attributed in an absolute way to any claims made by sciences, Quine was in effect arguing that "less false than other claims tested" could be a reasonable description of the semantic status of the best results of scientific research; it could be a useful goal for scientific methods.

Subsequent analyses of scientists' actual reasoning and social histories of science produced in the postwar period supported skepticism about the viability of Popper's "falsificationism." The history of science revealed frequent ad hoc assignments of error. Philosopher Paul Feyerabend polemically proposed that the new histories revealed that apparently "anything goes" in scientists' struggles to bring their observations and hypotheses into alignment with each other. Thomas S. Kuhn's studies of the history of science showed how the earlier linear model of the growth of scientific knowledge, whether inductively or deductively modeled, failed to capture the importance of conceptual shifts that left succeeding scientific theories if not completely incommensurable with each other, then at least not completely commensurable with each other in the way that the truth ideal required. Kuhn captured the difficulty with retaining either the truth ideal or an absolute falsity ideal in his observation that perhaps it would be better to say that science in general progresses away from falsity rather than that it progresses toward truth.[5] Kuhn's work often appears old-fashioned today, as it still assumes much of the traditional rationalist conceptual framework within which accounts of scientific progress were produced. For example, he refers to the important social and cultural influences on the growth of scientific knowledge as "irrational elements" in the history of science—a distinction that would be difficult to make in contemporary social studies of science. Nevertheless, in some ways his work remains pathbreaking and controversial outside the new science studies. He also understood that his suggestion that it would be useful to think of science as moving away from falsity rather than toward truth did not force one to a damaging relativism, for his historical account showed that scientific claims could reasonably, if only provisionally, be regarded as "less false" than those (and only those) against which they had been tested.

Recent Analyses

Since the 1960s, historians' arguments (including those of influential historians of philosophy) have continued to show that there is no evidence that either absolute truth or absolute falsity can ever be achieved in the sciences, and that there is a lot of evidence that they cannot. Moreover, the truth ideal leads to misunderstandings of how the best scientific beliefs have been achieved in the past and will be achieved in the future.

LOCAL FEATURES ARE PRODUCTIVE OF KNOWLEDGE

One problem is that the very elements of sciences and their practices responsible for historical, cultural specificity—"localness"—are also often responsible for successes. Moreover, there is no aspect of science that can be immunized from social and cultural influences. Such influences necessarily and often productively permeate its choices of problems, concepts, models of nature, research designs, techniques, instrumentation, collection and interpretation of data, ways in which scientists communicate with each other and other norms of scientific communities, and representation of results of research (mathematical or not). Of course, such social and cultural values and interests also affect patterns of dissemination, meanings, applications, technologies, and other consequences of scientific practices. Moreover, every element of scientific work contributes to the standards for scientific inquiry, and thus has philosophic consequences. Furthermore, culturally local resources do not just "influence" scientific practice, as the conventional language above suggests; they constitute its technical, cognitive core. So if "true" is taken to mean transcultural, then there is nothing that could be true in science.

THE CORRESPONDENCE THEORY OF TRUTH AND
THE UNITY OF SCIENCE: PROBLEMATIC

Moreover, these studies bring out another problem with the truth ideal. The prevailing conception of truth in the natural sciences has been a correspondence theory of truth: true scientific claims correspond to or are congruent with nature's order. This particular theory about truth cannot fully be understood in isolation from the rest of the epistemological model of which it is a constituent part. It is linked to a network of assumptions about the natural order, science, ideal standards of knowledge, the social order, and relations between such elements. This is a "metaweb" of belief that has developed and changed over time no less than scientific beliefs themselves.

The "unity of science" project has been one such constellation of assumptions within which the truth ideal has played a central role.[6] It forms part of the "political unconscious" of modern science that manages to persist even while those who assume it also are realistic about the extent of disunity among the existing sciences as well as about the positive value of this disunity. The unity project flourished in the first half of this century, but has been abandoned since World War II by much of the philosophical and historical tradition that developed it—in no small part due to the work of Quine, Feyerabend, Kuhn, and others writing a generation and more ago.

Nevertheless, its assumptions still shape leading philosophies of the social sciences, some scientists' conception of their mission, and popular thinking about the sciences. For instance, Edward O. Wilson, the influential author of *Sociobiology* and, more recently, *Consilience: The Unity of Knowledge,* has rigorously defended it. The persistence of this unity thesis, and the intensity with which it is held in the face of some four decades of by now widespread skepticism about it in the scholarly communities that specialize in the study of the history and present practices of science, suggests that belief in it fulfills a deeper psychic and cultural need for many scientists, as well as others, than mere counterevidence can hope to dislodge.

What is the relevance of the unity of science thesis to the truth ideal? The unity thesis overtly makes three claims: there exists just one unified world, one and only one possible true account of that world ("one truth"), and one unique science that can piece together the one account that will accurately reflect the one truth about that one world. The notion of absolute truth was central to this notion of the singularity of science. The viability of this still popular home for the truth ideal depends upon, first, the claim that the notion of absolute truth makes sense with respect to scientific claims and, second, the claim that the truth ideal can be a resource for the growth of knowledge. That is, apart from whether it is justifiable to regard science as producing claims that at least in principle can be known to be true (the first claim), we can ask if it is still useful to the growth of knowledge to assume that there is "one world" and one and only one science that could represent the unique correspondence relations between our claims about nature's order and nature's order itself. All three of these assumptions about the world, its true order, and science now appear to be either implausible or else no longer clearly and uncontroversially meaningful, and each seems to obstruct the growth of scientific knowledge.[7]

Ian Hacking had pointed out that the concept of one science has for different thinkers emphasized or sometimes blended two distinct ideas: singleness and integrated harmony. Singleness and integrated harmony have been weighted in different proportions in defenses of diverse kinds of metaphysical unity, methodological unity, and logical or "styles of reasoning" unity. Surveying diverse arguments for just metaphysical and methodological unities, Hacking identifies a metaphysical sentiment, three metaphysical theses, three practical precepts, and two logical maxims, each of which weights differently unity as singleness and as harmony. Moreover, there evidently are as many styles of scientific reasoning as there are of effective human reasoning more generally. As Hacking points out, A. C. Crombie identifies six important ones, each of which brings its own standards of adequacy.[8]

Indeed, it is now widely recognized that science cannot plausibly be understood as one single kind of thing at all. "There is no set of features peculiar to all the sciences, and possessed only by sciences. There is no necessary and sufficient conditions for being a science."[9] In the face of such disunity, many different kinds of techniques serve as unifiers, as Hacking refers to them. Mathematics is the earliest one recognized to have such a function. However, it turns out there is no "one thing" that is mathematics, since it, like the sciences to which it gives an appearance of unity, is a diverse collection of principles and practices, as historians of mathematics point out.[10] Numerous other such collections of unifiers are to be found among scientific instruments, techniques, attitudes, and in all those inventive strategies that occur in the "trading zones" within which scientists work to communicate across the diverse cultural and natural conditions that separate them.[11] Thus, the appearance of one truth, to which all of the sciences contribute their pieces, is created through innovative strategies that enable scientists to integrate more or less harmoniously their diverse understandings of nature's order(s) while still showing nature's role in generating the results of their research. Such accounts show scientific work to be much more difficult and creative than simplistic accounts of its truth ideal would indicate.

POSTCOLONIAL AND FEMINIST ACCOUNTS

These have also contributed to understanding the necessary "localness" of the best scientific claims. Postcolonial, anti-Eurocentric cross-cultural studies of other cultures' scientific and technological traditions emphasize the scientific value of the local resources that local cultures bring to their inquiries about how nature works, as discussed in earlier chapters. "High cultures," such as those of China or India, developed mathematical, scientific, and technological understandings that enabled them to flourish, many elements of which were later adopted into the sciences of the Global North. Such elements continue to arrive and attain adoption after centuries of mistaken rejection here. Moreover, even purportedly "underdeveloped" "simple" societies understand aspects of their natural environments that enable them to farm in fragile environments, prevent and cure diseases with local techniques and pharmacologies, and otherwise live well in their locations in nature's heterogeneous order.[12]

Of course, it is not that all belief identifiable as "local" is preferable to belief claimed to be universally valid—as was pointed out earlier. Obviously, many local beliefs lead their holders to nasty, brutish, and short lives. But the same can be said for some of the assumptions and beliefs of the sciences of the Global North. Northern scientific belief systems have contained systematic ignorance and error about environmental destruction, infectious diseases, human nutritional requirements, the effects of smoking, the effects of radiation, how to manage chronic pain, women's natures, the physical natures of nonwhite races, and many other topics. Politics as well as science—or, rather, scientific inquiry guided by politics—have been required to make space for the growth of knowledge on such topics. The low visibility of other cultures' scientific and technological achievements to Northerners has been due primarily to Eurocentric rhetorical and political practices, including the invention of "underdevelopment" by Northern nations and the international agencies in which their voices are predominant.[13]

Feminists have similarly charted distinctive patterns of systematic knowledge and ignorance in modern sciences that mark the latter as androcentric. Insofar as women have been assigned distinctive social activities, they will have distinctive interactions with their natural and social environments, and develop distinctive patterns of knowledge (and ignorance). Indeed, in every society they are assigned different activities in the household, the workplace, and in community life. Thus, women everywhere tend to have more knowledge about the bodies of infants, children, the elderly, and the sick, and in many parts of the world about agriculture, silviculture, animal husbandry, pharmacology, and any other work to which they are assigned.

Why All Sciences Must Be "Ethnosciences"

These three schools of post–World War II science and technology studies—Northern post-Kuhnian, multicultural and postcolonial, and feminist—all indicate that the constellation of human systems of knowledge about nature's order, including the sciences of the Global North today, is multicultural. There are many culturally distinctive local sciences that enable believers to succeed at a goodly number of projects while failing at others, projects that may be to some degree incompatible with other cultures' projects.

We can summarize why this necessarily occurs by recollecting five features of all knowledge systems (as I argued in an earlier chapter). First, they are developed for different locations in nature's order—for deserts or rain forests, for fertile plains or rocky islands, for environments shared with mosquitoes or AIDS viruses, for surviving on trips from Genoa to the Caribbean or from Cape Kennedy to the moon. And these environments themselves historically change through natural and social processes, offering ever new challenges to local knowledge systems.

Second, even in "the same" environment, different cultures have different interests in the world around them. Living on the shores of the Atlantic, one culture will be interested in fishing, another to use the sea as a coastal trade route, a third to retrieve oil and gas under the ocean floor, a fourth to use the sea as a garbage dump, a fifth to use it as a military route for submarines and torpedoes, yet another to desalinize its waters, and so forth. They will develop different patterns of systematic knowledge and systematic ignorance about oceans.

Third, they will have available to them different discursive traditions with different resources of metaphors, models, analogies, and narratives to be used to identify and explain the features of the world around them.

Fourth, each culture will have distinctive ways of producing knowledge, enabling it to detect some of nature's regularities while obscuring others. How the production of knowledge is organized is related to how cultures organize the rest of their activities—work, social relations, domestic life, and the like. How we interact with nature both enables and limits what we can know about it.

Finally, cultures are differently located in local and global power relations and will develop distinctive features in the preceding four categories according to the needs of such locations. Men and women, colonizers and the colonized, pursue different knowledge agendas. Thus, scientific and technological knowledge systems will always be local. Different patterns of knowledge cannot fit together like pieces of a jigsaw puzzle, as stronger forms of the unity of science thesis assumed, since some of these local patterns conflict with others. Even in the history of Western sciences one can reflect on the different representations of nature produced by thinking of it as an organism, a mechanism such as a clock, as a more complex mechanism such as a computer, or as a lifeboat or a spaceship. Each of these has played an important role in directing scientific practice to new knowledge and guiding sciences in revising assumptions when observations have failed to support hypotheses. Yet each is incompatible with the others in significant respects.

From the perspective of these kinds of reflections, the goal of only one true account of the natural world seems far too little to ask.

One Knower?

The unity of science thesis and its claim that there is just one science that can discover the one truth about nature also assumes that there is a distinctive universal human class—some distinctive group of humans—to whom the unique truth about the world could become evident.[14] However, as feminists and postcolonial thinkers have pointed out, this is no longer a plausible assumption for most of the world's peoples. This assumption has remained unarticulated and perhaps even unidentified by unity of science defenders as well as by most of their critics. Whom could such a group be? And how could any group achieve recognition in a fair way by other groups as having unique access to nature's truths?

Early modern scientists and philosophers, members of the new educated classes whose minds were trained to reflect the order of nature that God's mind had created (they assumed), were a select group. God's mind, human minds, and nature's order were assumed to be congruent with each other. Scientists and the educated classes that could see the truth and importance of scientific accounts represented the ideally human, the universal class that could learn to detect the one possible true account of nature's order. For nineteenth- and twentieth-century Marxists, it was the proletariat that represented the universal class. This class alone, since it was their labor that transformed nature into necessities for everyone's everyday life to exist, had the potential to understand natural and social orders that had been obscured by hegemonous bourgeois ideology, and thus to become the unique representative of human knowers. This class alone had the potential to detect the real relations of nature and social life beneath the distorted appearances produced by class society.

Some forms of feminism flirted with a similar kind of transvaluation of genders that considers the possibility that women are the uniquely human gender. If it made sense in sexist society to imagine men as the model of the uniquely human, then, they proposed, perhaps it is reasonable to consider how women's characteristics—their claimed altruism, peace loving, sensitivity to others' needs, centrality to maintaining community relations, or some other such virtue—provide uniquely valuable models of the human, capable of producing a better society and less distorted understandings of nature and social relations. Similarly, some African Americans have claimed that the suffering, compassion, or some other characteristic of African Americans under the horrible conditions of slavery and its aftermath uniquely equips them to understand natural and social orders in distinctive ways.

There are important insights behind such claims.[15] In the contemporary world of multicultural, postcolonial, and feminist politics and social theories, however, faith has declined in the possibility and desirability of such a universal class—whether Enlightenment, proletarian, womanly, or ethnically, geographically, or otherwise culturally distinctive. In the worlds of cultural diversity in which we all live at least to some extent, how could such a distinctively human universal class be identified? What group could democratically gain assent from others that it alone was capable of representing universal human interests, discursive resources, or ways of producing knowledge? Instead, we can see that different social groups, with their different historical experiences and inquiry practices, are indeed capable of making unique contributions to human understanding and knowledge. This seems a far preferable understanding of science than the grandiose and delusional claim that some one group's understanding of nature and social relations is uniquely entitled to represent all of human knowledge.

The Dysfunctional Truth Ideal

This brief account has identified a number of arguments against the desirability of the truth ideal that have appeared in three schools of science and technology studies developed in the past half century. My argument is not that there are no good uses for truth ideals in contemporary social life. Others may be able to show both their importance and that such ideals in social relations can meet the kinds of objections that have been raised about the usefulness of such an ideal in the sciences. Rather, the argument here is that main tendencies in several recent generations of philosophy and history of science, including feminist and postcolonial studies during this same period, fail to support the assumption that the growth of either knowledge or of democracy is advanced by claims to the truth of science's results of research.

This issue about the "politics of truth" should not be ignored. In light of the importance of historically distinctive cultural elements to the growth of scientific knowledge, any truth ideal that insists on eliminating cultural elements from science, let alone on refusing to critically examine the cultural elements in one's own society's sciences, and especially those that have been identified by others, works against the nourishment of democratic social relations to which the institutions of science are supposed to be committed. Insistence on or assumption of such a truth ideal advances inequality. In turn, the reduction of democratic social relations destroys resources, most notably a critical attitude toward dominant beliefs and assumptions, that all cultures have developed to aid them in understanding natural and social relations. So the truth ideal's antidemocratic and antiscientific tendencies support each other.

The fact that these analyses have developed during this particular historical period is significant, for whereas the cold war that began in the past half century brought an immense increase in Western governments' investment in scientific and technological research projects that were thought to advance Western military interests, the end of World War II also marked the beginnings of a rising chorus of complaints about the costs of modern sciences and their ideals of truth, objectivity, and rationality for politically vulnerable groups and for democratic social relations. This theme of the past half century's science studies—Northern, Southern, and feminist—enables us to see that the cultural resources sciences use are toolboxes as well as prison houses for the growth of knowledge, enabling the production of patterns both of systematic knowledge and of systematic ignorance. Each historical and cultural moment will bring its own fresh possibilities for understanding yet other aspects of nature's order, though the useful urge to integrate disparate patterns of knowledge will always result in the loss as well as the expansion of knowledge. The kinds of cognitive diversity generated by culturally local knowledge projects are a great resource of our species. The truth ideal shrinks rather than expands such resources.

What can be substituted for the truth ideal? Evidently, only something much less inspiring and impressive to aspiring young scientists and to those groups to whom have been distributed a disproportionate share of the benefits of the sciences of the Global North. Powerful groups do not want to give up the truth ideal since to do so challenges their right to decide how their society is to interact with nature and how social relations are to be organized. And in societies where the truth ideal is wielded by the powerful, resistant groups often think that they, too, must be able to claim that their perceptions of nature and social relations are true. For example, feminists must still struggle mightily to get recognition from dominant institutions of women's actual living conditions—of the double day of work, of women's lack of human rights, of their routine subjection to sexual violence, and of their lack of access to the kinds of nutritional, health, and economic resources available to their brothers (when it is). Victims of racial discrimination continue to have to fight for similar recognitions. It can seem crucial to these groups to be able to claim that their analyses are objectively true, not just their own particular perceptions.

Whatever may be the kinds of rhetoric that the political struggles of the moment seem to require of the most economically, politically, and socially vulnerable of the world's citizens, it is equally important to insist that the dominant groups find a more widely beneficial substitute for their rhetoric about the truth ideal. Fidelity to what inquiry can actually achieve can be a reasonable standard here. Thus, we can aim for the provisionally least false of all and only the hypotheses already tested. Adopting this standard would promote sciences that can avoid advancing inequality and instead contribute toward democratic goals. As we have seen, such sciences would also be more cognitively competent than can be those that cling to the truth ideal.

9

Does the Threat of Relativism
Deserve a Panic?

Could there be one transcultural standard for evaluating all beliefs about nature and social relations? Modern Western sciences and their philosophies have assumed that their standards are the one and only ones to so qualify. However, widespread doubts have arisen that there could be any such transcultural standard, as well as that there could be a universally legitimated process that would show that modern Western sciences and their philosophies come the closest to such an ideal.

Feminists have argued that the conventional standards for maximizing objectivity have evidently been too weak to permit the identification of the sexist and androcentric values and interests that have permeated scientific research. Prevailing standards for the rationality of science and for good method have met similar criticisms. These standards are too weak to function effectively when particular values and interests are widely shared within a scientific community, and especially when they are also the dominant beliefs of the surrounding society.[1] Thus, these standards for good science are themselves discriminatory since they favor the values and interests of culturally distinctive groups, namely, those that fund, sponsor, and direct scientific research and that, three decades after they first appeared, still resist incorporating feminist insights into their thinking about nature and science. Scientific standards, like the scientific claims they justify, are always socially situated.

Other groups have raised similar objections to the kinds of rationally justifiable standards conventionally favored in scientific work and its philosophies. Such objections appear in poststructuralist criticisms of Enlightenment assumptions,[2] in postpositivist philosophy's and social studies of science's focus on the cultural features of "good science,"[3] and in the criticisms of Eurocentric assumptions raised in multicultural and postcolonial science studies.[4]

Yet many find it disturbing that these criticisms seem to abandon or even reject the hope that a single transcultural standard exists that can be used in evaluating competing beliefs with respect to ontological facts about the world, epistemic rules or principles for evaluating knowledge claims, and methodological norms. According to this line of thought, without one such standard, we have no way to rationally justify any particular standards at all when faced with conflicting knowledge claims; each is simply preferred for cultural or political reasons. We are doomed to a damaging epistemological or judgmental relativism such that anyone's opinion, regardless of whether *we* think it rationally justifiable, must be regarded as just as good as any other, including our own.

There are many misunderstandings of relativism. Let me emphasize here the difference between sociological or historical relativism, on the one hand, and epistemic relativism, on the other hand. The former asserts only the fact that different groups have different standards. This fact is obvious, and everyone should be able to agree to it. Historical or sociological relativism is compatible with either an absolutist or relativist epistemological position. The epistemological absolutist says that only one such standard can be the right one, and the epistemological relativist says that since there is not and cannot be such a standard, all must be regarded as equally right or true. One would think that studies of the natural sciences would be loath to adopt an epistemological relativist position, yet several well-known sociologists of science have done so.[5] Nevertheless, epistemological relativism is a minority position in science studies. An absolutist position, supported by antirelativist fears, is probably the most widely held position in the natural sciences and much of the social sciences, as well as in conventional philosophy of science.

To be sure, the possibility of epistemic relativism is indeed a disturbing perception. We need rationally justifiable reasons for medical, health, nutritional, environmental, social policy, and a host of other decisions with which we are faced on a daily basis. We need to be able to convince those who disagree with us, not just those who are already on our side, in any particular dispute.

However, there is no good reason to think we are stuck with a damaging relativism just because we give up the idea that our own Western scientific standards are the only justifiable ones. Indeed, the extreme fear of relativism—the "antirelativism" obsession—keeps us committed to empirically, theoretically, and politically insupportable positions. We would do better to get clear just what the real problem with antirelativism is, as anthropologist Clifford Geertz puts the point. "The objection to antirelativism is not that it rejects an it's-all-how-you-look-at-it approach to knowledge or a when-in-Rome approach to morality, but that it imagines that they can only be defeated by placing morality beyond culture and knowledge beyond both."[6] Geertz defends not relativism, but rather resisting the lure of antirelativism. We can give up the illusion that science can be beyond culture and morality without either adopting a damaging cognitive or epistemological relativism or getting lured into an antirelativist obsession.

Three Responses to Challenges to a Single Standard

But let us return for the moment to the criticisms of the single-standard ideal and how it is perceived with respect to the objectivity standard. Three such responses to these criticisms are widely visible. One is to ignore them and hope they will go away. "I prefer weak objectivity," one correspondent informed me, referring to the traditional standard that requires scientific beliefs and practices to be value neutral if they are to achieve objectivity. A second response is to argue for abandoning the language of objectivity and conventional procedures for maximizing it and, instead, to appreciate the virtues of subjectivism or relativism or both. A long history of anthropologists and even some contemporary philosophers have recommended relativism's virtues. For example, philosopher Paul Feyerabend has argued for them. Feminist Lorraine Code has framed her epistemological arguments in terms of advancing relativism. And cultural critic Fredric Jameson has admired standpoint theory in pointing out that one of its distinctive features is its "principled relativism."[7] A third response has been to insist on the importance of understanding how sciences are always constituted within culturally distinctive conceptual schemes and practices, but also on the importance of strengthening both the existing ideal and procedures for achieving objectivity. This has been the direction taken by many feminist and postcolonial critics, for example, Donna Haraway, Evelyn Fox Keller, and myself.[8] The social justice movements need "to have *simultaneously* an account of radical historical contingency for all knowledge claims and knowing subjects, a critical practice for recognizing our own 'semiotic technologies' for making meanings, *and* a no-nonsense commitment to faithful accounts of a 'real' world." The sciences need this too, though they are resistant to recognizing that fact. Because the sciences and their philosophies lack the first two components in Haraway's description, they have remained "epistemologically underdeveloped."[9]

The fear of relativism seems odd, also, when we consider that in daily life we are perfectly able to produce what most people, including law courts, regard as rational justifications for our knowledge claims. We do not think that such claims are absolutely true, under any conditions, now and forever; they are always revisable if additional evidence or a useful new conceptual framework appears. We are used to the idea that we may reasonably change our mind about facts, even about what we witnessed with our own eyes—indeed, that it is rationally necessary for us to do so when we come across not only additional facts but also illuminating new conceptual frameworks and practices. We know that things are not always what they seem to be. Yet we believe that we have good evidence and a sound argument that justify holding such beliefs here and now and rejecting other competing beliefs. Whether we are right or wrong to do so in particular cases, we routinely and confidently take such positions with respect to health matters, legal issues, and the everyday choices we must make.[10] The arguments between absolutists and relativists seem to float free of such everyday experiences and the ways we think about them.

If one agrees that even one's most defensible beliefs about the world around us are shaped by culture (as those new movements claim), has one forfeited any possibility of legitimate rational justification for belief preferences? If one admits that knowledge is always constituted only within particular historical discourses, has one forfeited the possibility of establishing empirical support for one's beliefs through referring to anything beyond discourse? If even one's ontological, epistemic, and methodological standards for sorting beliefs have an "integrity" with their historical era (as Thomas Kuhn put the point four decades ago), is it possible to make objective, reliable, and fair decisions between conflicting theoretical choices in the natural and social sciences? Even more radically, what would it mean to say that the overtly culturally embedded beliefs that other non-Western cultures and feminist projects have about nature's order or about social relations can be as universally valid, as objective as—or even more valid and objective than—modern Western scientific beliefs that purportedly are culturally neutral?

The antirelativism versus relativism literature is huge; it occurs in many disciplines, but especially in anthropology and philosophy, which have historically taken responsibility for identifying, respectively, the irrational and the rational in human thought and action.[11] These focus on the limits of translation, the adequacy of the familiar dualism of conceptual scheme versus content, the coherence of the very idea of a conceptual scheme, the problems with representationalist theories of scientific knowledge, and a host of other issues. There have been compelling arguments from many quarters for rejecting both the absolutist, obsessively antirelativist position and the damaging relativism with which a few people say they are willing to live. In one way or another these arguments draw attention to how we can have good evidence for a belief, that is, the best possible in some particular context, yet not need or want to hold that this evidence or the belief must be held as transculturally compelling. Here I show just one such way of escaping the antirelativist and relativist positions. In doing so I bring together issues raised in earlier chapters.

Two Faulty Assumptions

Absolutist antirelativism shares two faulty beliefs with the relativist position it so fears. One is that words and the world can and should be completely separable or distinguishable from each other if we are to have rationally justifiable standards for belief. If what we take to be objects and processes in the world cannot be separated from the discursive frameworks and practices through which we learn to identify and think about them, then there can be no rational way to choose between knowledge claims from different such frameworks.[12]

To be sure, the ability to separate words from the world, to think abstractly, is a powerful tool. The issue is not whether we should give up abstract thought and simply interact with the concrete objects here and now before us. Rather, the issue is whether one can sustain the ideal of rational, objective belief choices in a world where words and the things to which they refer have proved in principle not as isolatable, or even as desirable to isolate, as modern philosophies of science presumed. Should one try to sustain an ideal of rational, objective choices between conflicting beliefs? They answer here is an unabashed yes—not, to be sure, a radical position in itself.

However, such a goal is achievable only by abandoning a second basic assumption shared by absolutists and relativists, namely, that what counts as science or knowledge is fundamentally a set of words or representations of the world. One expression of this assumption is that real science and true knowledge must be formulated as sentences or statements about the world's natural or social (or both) nature and workings, and that such representations, at least in principle, must perfectly correspond to the way the world is.

This representational notion of science has always been held alongside a quite different one: that science is a distinctive method of inquiry, a set of practices. It is an effective way of going about figuring out how natural and social environments work. This second notion finds a home in sciences' concern with designing and conducting research in ways that are fair to the way the world is (to the data), and fair to the severest criticisms of one's inquiry methods. It also is the basis for the liberal democratic insistence on the importance of fair procedures in the law and in public policy. The "practice" notion of science was emphasized by Marx in his observation that science is fundamentally about interventions in nature. It has been reworked and reimagined in the past half century by Thomas Kuhn in his discussion of a research "paradigm" as an exemplar that precedes a new way of theorizing nature's order, in Michel Foucault's discussion of the importance of the founding of the modern prison and clinic as models for methods of observation for the emerging social sciences, in Ian Hacking's eloquent account of the unjustly overlooked importance of intervention to scientific work, and in the development of such insights in Joseph Rouse's work and, very recently, in Karen Barad's.[13] Philosophers, sociologists, ethnographers, and historians of science have focused on scientific practice in their attempt to replace excessively idealized representations of scientific work with ones "naturalized" through more careful empirical study of actual practices. And comparative ethnographies of other cultures' knowledge projects as well as postcolonial and feminist science and technology studies have carefully examined inquiry practices in a far broader range of cultures and subcultures than the idealized versions of "real science" usually considered relevant.[14] Thus, the account here brings together insights from diverse sources that form an emerging tendency in philosophies of science to look at how sciences create knowledge through patterns of practice.

My point here is that it is difficult for epistemic relativism to arise if science is thought of as a set of distinctive practices that generate, among other products, representations of nature and social relations such as accurate descriptions and perhaps even causal laws. One can always ask of a practice what its goal is, and then judge the desirability of the goal, the adequacy of the practice in reaching that goal, and the extra effects of the practice in addition to reaching the goal. All of these decision processes can be controversial, but at least these controversies are more relevant to how a particular set of scientific practices affects people's lives than generalized debates between antirelativist absolutists and relativists. Thus, the proposal here is that we can abandon epistemic absolutism and its goal of articulating the one true story of the natural and social order without giving up rational grounds for our best beliefs. Those grounds are both practical-methodological and ethical-political.

One final perception here. It is useful to note that the very different social movements whose research projects generated such reactions against epistemic, ontological, and methodological forms of both absolutism and relativism began to emerge at about the same time—in the 1960s and early 1970s. Moreover, intentionally or not, all of these epistemology and science movements arose alongside significant changes in the global social formation that have brought into existence or made visible a world importantly different from the one imagined in early-twentieth-century mainstream philosophies of science.[15] Among such changes have been the end of formal European colonial rule; new economic, political, and social roles for women; rising skepticism about the allegiances of natural and social sciences to militarism, capitalism, white supremacy, and male supremacy; globalization of the economy and the shift from industrial production to the production and management of information; the rise of new identity movements around the world; and the increased dissemination of liberal democracies and rising resistance to them from both conservative and progressive groups around the world.[16]

My point is that evidently the world changed in the past half century in ways that unsettled conventional certainties about how knowledge projects are, can be, and should be situated in the world. Changes in epistemology and philosophy of science, however else one might characterize them, have been attempts to update or revise older models of how and for what ends knowledge is produced because these models are perceived as out of touch with what we now know about the history and present practices of natural and social science inquiry. Philosophies are always engaged with their eras' problematics. Giving up these two aspects of conventional philosophies of sciences has other consequences.

Moving Past the Modernist Dream:
One Postpositivist Approach

The beginnings of the disillusion with a representationalist conception of "real science," and of the shift to a focus on practice, can be found in the philosophy of science of the 1950s and 1960s. For example, W. V. O. Quine argued that empirical observations and the theories they are supposed to test can never achieve the independence from each other that both the inductivist and the deductivist "logics of scientific inquiry" required. Nor are analytic and synthetic statements as discrete as had been thought. Our scientific and everyday beliefs form a network such that when observations conflict with a favored theory, scientists might even find it more reasonable to revise a law of logic rather than to give up any other empirical or theoretical belief. Obviously, this would be an extremely drastic and rare choice, but it is not out of the question, as it would be for the older philosophies of science.[17]

Then Kuhn argued that the new social histories of science emerging in his day pointed to the way moments in the history of the very best scientific research had an "integrity" with their historic era. That is, what Kuhn still referred to as "irrational" cultural factors regularly played crucial roles in distinctive ways of advancing the growth of knowledge. The conventional logical positivist attempt to show that the greatest achievements of Western modern sciences were entirely a result of purely rational practices (plus nature's order) was a mistake: the "paradigm shifts" that periodically reorganized and energized scientific fields were not explicable in the conventional purely rational terms. Kuhn's paradigms were not new theories, he insisted; they were practical exemplars of scientific research—research practices—that subsequent researchers used as a model for their own work. Any retheorization of nature's regularities could occur only later. Meanwhile, a similar point was made by Foucault's focus on how the organization and practices of the nineteenth-century clinic and prison modeled social research for the subsequently emerging social science disciplines, as noted above. The practices necessarily preceded the theorizations. His point was that knowledge and power were inextricably linked through particular institutional practices. The bodies of patients and prisoners were forced to release, give up, "confess," the information desired by those who designed the clinic and prison, who would in turn use this information to control additional groups that were already socially disempowered (the diseased, the criminal).[18] The social sciences learned to exercise something like this kind of power to produce their results of research.

Thus, from a number of sources has appeared the argument that scientific practices (and nature itself) produce many phenomena: new kinds of laboratories, techniques of inquiry, networks of researchers and their support staffs, newly visible—perhaps even newly created—things and processes (geologic plates, retroviruses, double helices), demands for new kinds of schooling, new rhetorics of scientific progress, demands for funding, and, yes, sets of sentences that it is hoped represent nature's order accurately enough for the purposes at hand.[19]

Such an understanding of the nature of science requires a different philosophy of science. Developing such a philosophy is an important project for the prodemocratic new social movements. These movements need to be able to justify the rationality and objectivity of their explicitly politically committed inquiry practices. Such practices are intended to reveal how implicitly politically committed (intentionally or not) are androcentric, Eurocentric, and class-exploitative conceptual frameworks of social and natural sciences that are presumed value neutral. The prodemocratic social movements do not take such accounts to be merely subjective opinions. That is, they need to be able to show that research that appears to be value neutral in fact is not, and also that some social values and interests are productive both of knowledge and of social justice—both of which have historically been goals of objective scientific inquiry. One can identify central foci of the still emerging network of postpositivist philosophies of science. (These nodes are not completely independent of each other, of course.)

NETWORKS OF BELIEF

The kinds of beliefs we need to get around in the world cannot in principle be completely isolated from each other. Instead, they must form a dense network that links empirical, theoretical, logical, political, economic, social, psychological, and practical everyday assumptions and beliefs.

MULTIPLE, DISCORDANT, SCIENCES

Yet there cannot be only one such network; there must be many both interlocked and conflicting ones. Science is not singular, either in practice or in principle, as the strong form of the unity of science thesis held.[20] Scientific elements are not even actually or ideally always "harmonious," as the weaker form of the thesis proposed. This multiplicity and discordance are both problematic and fortuitous. It can be useful to link together different kinds of observations and discordant theories as, for example, Darwin did. Discordances between knowledge systems characterize the conceptual shifts ("paradigm changes") through which scientists reorganize most of a field of existing data into illuminating new patterns. Moreover, such discordances mark the valuably distinctive positional, interest, discursive, and organizational resources that different fields, disciplines, and cultures bring to their attempts to understand their allotted or chosen environments.

CULTURE AND POLITICS CAN BE PRODUCTIVE OF
THE GROWTH OF KNOWLEDGE

Thus, distinctive political and cultural resources—interests, values, and culturally familiar discursive resources such as analogies, narratives, and models—can sometimes advance the growth of knowledge. Of course, these can also be obstacles to knowledge, but that is not their only possibility. Cultures are "toolboxes," not just "prison houses," for the production of knowledge.[21] This is another way of saying that words and things cannot be completely separated and distinguished from each other such that there are facts about the things that can be used to judge the absolute adequacy of our theories (words) about them—one of Quine's and Kuhn's insights. The research of the prodemocratic social movements has always been engaged, committed to improving the conditions of oppressed lives—though always also resisting the tendency to turn wishes into purported facts. So, too, has mainstream mission-directed research, whether intended to fight diseases or improve the accuracy of artillery. Politics has always been part of scientific method, one could say.

MIND-INDEPENDENT REALITY: ALWAYS OUT OF SIGHT

Yes, there is a mind-independent reality—or, perhaps, many of them. It was there before humans appeared and will persist when we become extinct. Yet it cannot be that our best knowledge claims are those that uniquely correspond to it. Many theories are consistent with nature's order, but no single one could be uniquely congruent with it.[22] Instead, all scientific method ever promised was to identify the best hypothesis, for the moment, of all and only those tested—not the one and only one that could reign eternally unchallengeable. Fortunately, there are always many, many more hypotheses not yet tested, including the quadrillions not yet even formulated, which offer the permanent promise of new and surprising insights into nature's order and social relations. Thus, truth in the standard "correspondence" sense of the word prevalent in the sciences—a perfect match between words and their referents—is not a useful goal for scientific practices, or at least for a philosophic account of them. Even the positivists understood that only the claims of logic and religious or other forms of dogma, as they put the point, both immune to empirical tests, could legitimately claim truth. "Temporarily least false of all and only those considered" is the most that scientific method could promise.[23]

SCIENCES AS CULTURALLY EMBEDDED PATTERNS OF PRACTICE

Sciences are most usefully conceptualized not as sets of transcultural authoritative statements or representations, such as the laws of nature, but rather as practices embedded in cultural histories, interests, and values. Indeed, cultures store within preexisting cultural frameworks the kinds of information about the world that are most important and useful to them—beliefs about how the Christian God made order out of chaos creating a Garden of Eden for his chosen people, about how the earth is like a spaceship or lifeboat such that its environment must be carefully monitored, and so forth. We learn the world and the pattern of practice (including words) together.[24]

DECENTERED COLLECTIVE SUBJECTS OF SCIENCES

Thus, the "subject" of science and knowledge—the collective speaker—cannot be the centered, transcultural rational knower of Enlightenment philosophy, but must instead be human cultures in all of their systematic ways of successfully interacting with the worlds around them. Such an understanding can trigger absolutist fears of relativism, yet the argument here has been that such a fear is unnecessary. If one values both Buddhism and the management of chronic pain, it can be rational to seek acupuncture, a practice that is meaningful in the context of Buddhism and also independently effective in controlling chronic pain. However, if one values the management of chronic pain and only those belief systems free of religious assumptions, until recently one would have had to choose between commitment to a set of apparently effective practices that for many people carry religious assumptions and Western biomedicine that purportedly distances its accounts from any apparently religious assumptions but had not developed practices effective at managing chronic pain. Now that Western biomedicine has explained the effectiveness of acupuncture in its own terms (whether or not they are free of religious assumptions), one no longer has to make such a choice.[25]

Abandoning Absolutism and Relativism

Relativism was always the invention, the nightmare, of the absolutist, even if often appropriated by others for either progressive or regressive scientific or political projects.[26] Here I have been outlining how we can still have a world "out there" beyond our historically local discourses, but we cannot have one that authoritatively chooses for us which knowledge claims to believe. Observations always must be interpreted within socially meaningful frameworks. We can still have justifiable standards for defending our beliefs and practices, but we cannot have ones that do so apart from our investments in the perceived strengths of the social values and interests that make conceptual frameworks meaningful. We can still have sciences that illuminate nature's order and social relations for us, but we cannot have ones that uniquely do so for everyone and thus are reasonably regarded as effective and desirable for every culture and its historically local projects. We can have modern Western sciences that can be the very best that we can produce given the cultural and environmental resources available to us and the goals for which we have designed our sciences, but we cannot have ones that are the best we could produce for any goals we might have or, consequently, that we can confidently depend upon to provide all that we might desire to know in the future.

Indeed, we can be confident that the sciences thought to have advanced the forms of democratic social relations envisioned by the Enlightenment and its heirs today are most likely not the ones that work well to advance the new forms of democratic social relations for which feminisms, multiculturalisms, and postcolonialisms yearn today. Nor can they engage effectively with the recently appearing new ways of producing scientific knowledge. Historically distinctive social relations and their favored forms of knowledge seeking co-constitute each other. The rise of new social values, interests, and the relations they direct requires inquiry practices and principles that can support and in turn be supported by these new forms of, we hope, democratic social relations. Our methodological and epistemological choices are always also ethical and political choices.

Indeed, the choices shaping this very account, including those invisible to its author, can be only one possible response to documented effects of sciences in the world today and contemporary yearnings for a better world. Philosophies of sciences, too, must be multiple. The adequacy of this one will be measured by how it and others like it are used to shape future social practices.

Notes

Introduction

1. Is the standard Western conception of democracy part of the problem here? Liberal democracy, a capitalist economic system, and modern sciences all emerged and developed together as an interlinked social formation. They share common conceptual features, as will be discussed in the chapters that follow. Political philosophers have argued that liberal democracy has been ill-equipped to engage productively with differences between social groups. See, for example, Seyla Benhabib, ed., *Democracy and Difference: Contesting the Boundaries of the Political* (Princeton: Princeton University Press, 1996). Moreover, neoliberal democracy is promoted globally primarily to keep the world safe for Western profiteering, they argue. See, for example, Stephen Gill, "Globalisation, Market Civilisation, and Disciplinary Neoliberalism," *Millennium: Journal of International Studies* 24.3 (1995): 399–423; and William I. Robinson, *Promoting Polyarchy: Globalization, U.S. Intervention, and Hegemony* (New York: Cambridge University Press, 1996).

2. Charles Schwartz provides a list of common defenses scientists make of the social neutrality of their own military research in "Political Structuring of the Institutions of Science," in *Naked Science: Anthropological Inquiry into Boundaries, Power, and Knowledge*, ed. Laura Nader, 148–59 (New York: Routledge, 1996). For example, "Research is essential so that we know what threatening weapons are possible"; "I only work on defensive weapons, not offensive ones"; "If I don't do this work on weapons, someone else will"; "I am just a scientist doing my job; I stay out of politics." These, plus the additional eleven such defenses Schwartz identifies, make a great discussion starter. As he points out, "Each has a core of truth, but also a serious shortcoming" (157).

3. I borrow the term here from Hannah Arendt, *Eichmann in Jerusalem: A Report on the Banality of Evil* (New York: Viking Press, 1963).

4. See Paul R. Gross and Norman Levitt, *Higher Superstition: The Academic Left and Its Quarrels with Science* (Baltimore: Johns Hopkins University Press, 1994); Paul R. Gross, Norman Levitt, and Martin W. Lewis, eds., *The Flight from Science and Reason* (New York: New York Academy of Sciences, 1997); Andrew Ross, ed., *The Science Wars* (Durham: Duke University Press, 1996); and Alan Sokal and Jean Bricmont, *Fashionable Nonsense: Postmodern Intellectuals' Abuse of Science* (New York: Picador USA, 1998).

5. Fredric Jameson, "On Interpretation: Literature as a Socially Symbolic Act," in *The Political Unconscious: Narrative as a Socially Symbolic Act* (Ithaca: Cornell University Press, 1981), excerpted in Michael Hardt and Kathi Weeks, eds., *The Jameson Reader* (Oxford: Blackwell, 2000), 35.

6. See, for example, Meera Nanda, *Prophets Facing Backward: Postmodern Critiques of Science and Hindu Nationalism in India* (New Brunswick, N.J.: Rutgers University Press, 2004).

7. See, for example, Peter Galison and David J. Stump, eds., *The Disunity of Science: Boundaries, Contexts, and Power* (Stanford: Stanford University Press, 1996).

8. Two reviews of the issues this literature raises for Western histories and philosophies of science may be found in Sandra Harding, *Is Science Multicultural? Postcolonialisms, Feminisms, and Epistemologies* (Bloomington: Indiana University Press, 1998); and David J. Hess, *Science and Technology in a Multicultural World: The Cultural Politics of Facts and Artifacts* (New York: Columbia University Press, 1995). See also Helaine Selin, ed., *Encyclopedia of the History of Science, Technology, and Medicine in Non-Western Cultures* (Dordrecht: Kluwer, 1997); and the *Indigenous Knowledge and Development Monitor* (http://www.nuffic.ni/ciran/ikdm.html).

9. One place "the nature of science" is currently much debated is in science education contexts. See, for example, *Multiculturalism and Science Education,* a special issue of *Science Education* 85.1 (2001).

10. For example, Donna Haraway points to the puzzlement of Western primatologists upon discovering that some of the pathbreaking Japanese primatology research involved human feeding of "wild" monkeys. See "The Bio-politics of a Multicultural Field," in *Primate Visions: Gender, Race, and Nature in the World of Modern Science* (New York: Routledge, 1989).

11. See, for example, Karin Knorr-Cetina, *The Manufacture of Knowledge: An Essay on the Constructivist and Contextual Nature of Science* (New York: Pergamon, 1981); and David Bloor, *Knowledge and Social Imagery* (London: Routledge and Kegan Paul, 1977).

12. Thomas S. Kuhn, *The Structure of Scientific Revolutions,* 2d ed. (Chicago: University of Chicago Press, 1970), 1.

13. See, for example, Manuel Castells, *The Information Age: Economy, Society, and Culture,* vols. 1–3 (Oxford: Blackwell, 1996–2000).

14. See Michael Gibbons, Camille Limoges, Helga Nowotny, Simon Schwartzman, Peter Scott, and Martin Trow, *The New Production of Knowledge: The Dynamics of Science and Research in Contemporary Societies* (Thousand Oaks, Calif.: Sage, 1994).

15. Ulrich Beck, *Risk Society: Towards a New Modernity* (London: Sage, 1992); Beck,

World Risk Society (Oxford: Blackwell, 1999); Anthony Giddens, *The Consequences of Modernity* (Stanford: Stanford University Press, 1990).

16. See Nader, *Naked Science.*

17. See, for example, their pamphlet *Modern Science in Crisis: A Third World Response* (Penang, Malaysia: Third World Network, 1988) and their journal, *Third World Resurgence.*

18. See Wolfgang Sachs, ed., *The Development Dictionary: A Guide to Knowledge as Power* (Atlantic Highlands, N.J.: Zed Press, 1992); or Arturo Escobar, *Encountering Development: The Making and Unmaking of the Third World* (Princeton: Princeton University Press, 1995).

19. See Edward Said, *Orientalism* (New York: Pantheon, 1978).

Chapter 1. Thinking about Race and Science

1. See David Wellman, "Prejudiced People Are Not the Only Racists in America," in his *Portraits of White Racism* (New York: Cambridge University Press, 1977).

2. For a description and analysis of the causes of continuing racial stratification in the United States, see Michael Omi and Howard Winant, *Racial Formation in the United States* (New York: Routledge, 1994).

3. Stephen Jay Gould discusses this phenomenon in *The Mismeasure of Man* (New York: W. W. Norton, 1981). See also Omi and Winant's discussion in *Racial Formation.*

4. See Gould, *The Mismeasure of Man;* and Richard C. Lewontin, Steven Rose, and Leon J. Kamin, *Not in Our Genes* (New York: Pantheon, 1984).

5. See, for example, Robert Proctor, *Racial Hygiene: Medicine under the Nazis* (Cambridge: Harvard University Press, 1988); James H. Jones, *Bad Blood: The Tuskegee Syphilis Experiment* (New York: Free Press, 1981); Rita Arditti, Renate Duelli-Klein, and Shelly Minden, eds., *Test-Tube Women: What Future for Motherhood?* (Boston: Pandora Press, 1984); and Robert Bullard, *Dumping in Dixie: Race, Class, and Environmental Quality* (Boulder: Westview Press, 2000).

6. *Minorities in Science: The Pipeline Problem,* a special issue of *Science* 258 (1992); Kenneth R. Manning, *Black Apollo of Science: The Life of Ernest Everett Just* (Oxford: Oxford University Press, 1983); Darlene Clark Hine, "Co-laborers in the Work of the Lord: Nineteenth-Century Black Women Physicians," in *"Send Us a Lady Physician": Women Doctors in America, 1835–1920,* ed. Ruth J. Abram (New York: W. W. Norton, 1985).

7. See Gould, *The Mismeasure of Man;* Winthrop Jordan, *White Over Black: American Attitudes toward the Negro, 1550–1812* (Chapel Hill: University of North Carolina Press, 1968); Proctor, *Racial Hygiene;* and Nancy Leys Stepan, *The Idea of Race in Science: Great Britain, 1800–1960* (London: Macmillan, 1982).

8. Proctor, *Racial Hygiene.*

9. Gould, *The Mismeasure of Man;* A. R. Jensen, *Bias in Mental Testing* (New York: Free Press, 1980); Lewontin, Rose, and Kamin, *Not in Our Genes;* Edward O. Wilson, *Sociobiology: The New Synthesis* (Cambridge: Harvard University Press, 1975).

10. Evelynn Hammonds, *The Logic of Difference: A History of Race in Science and Medicine in the United States* (forthcoming).

11. See the essays in pt. 2 of Sandra Harding, ed., *The "Racial" Economy of Science: Toward a Democratic Future* (Bloomington: Indiana University Press, 1993).

12. See especially Anne Fausto-Sterling, "Refashioning Race: DNA and the Politics of Health Care," *Differences: A Journal of Feminist Cultural Studies* 15.3 (2004): 1–37, and a forthcoming article in *Signs: Journal of Women in Culture and Society* (2005); and Hammonds, *Logic of Difference*. See also Fausto-Sterling, "The Bare Bones of Sex: Sex and Gender," *Signs: Journal of Women in Culture and Society* 30.2 (2005): 1491–1528.

13. Fausto-Sterling, "Refashioning Race."

14. Ibid, 37.

15. See Donna Haraway, *Primate Visions: Gender, Race, and Nature in the World of Modern Science* (New York: Routledge, 1989); *Simians, Cyborgs, and Women: The Reinvention of Nature* (New York: Routledge, 1991); and *Modest_Witness@Second_Millennium.FemaleMan_Meets_Oncomouse: Feminism and Technoscience* (New York: Routledge, 1997).

16. Sander L. Gilman, *Difference and Pathology: Stereotypes of Sexuality, Race, and Madness* (Ithaca: Cornell University Press, 1985); Hammonds, *Logic of Difference*; Haraway, *Primate Visions*; Haraway, *Simians, Cyborgs, and Women*; Haraway, *Modest_Witness@Second_Millennium*; Nancy Leys Stepan, "Race and Gender: The Role of Analogy in Science," *Isis* 77 (1986).

17. Rosi Braidotti, E. Charkiewicz, Sabina Hausler, and Saskia Wieringa, *Women, the Environment, and Sustainable Development* (Atlantic Highlands, N.J.: Zed Press, 1994); Daniel R. Headrick, ed., *The Tools of Empire: Technology and European Imperialism in the Nineteenth Century* (New York: Oxford University Press, 1981); Jones, *Bad Blood*; Proctor, *Racial Hygiene*; Wolfgang Sachs, ed., *The Development Dictionary: A Guide to Knowledge as Power* (Atlantic Highlands, N.J.: Zed Press, 1992); Vandana Shiva, *Staying Alive: Women, Ecology, and Development* (London: Zed Press, 1989).

18. See Harding, *"Racial" Economy of Science.*

19. Manning, *Black Apollo*; Willie Pearson Jr., *Black Scientists, White Society, and Colorless Science: A Study of Universalism in American Science* (Milwood, N.Y.: Associated Faculty Press, 1985); *Minorities in Science: The Pipeline Problem*, a special issue of *Science* (1992).

20. Nancy Brickhouse, "Embodying Science: A Feminist Perspective on Learning," *Journal of Research in Science Teaching* 38.3 (2001): 282–95.

21. Manning, *Black Apollo*; *Minorities in Science: The Pipeline Problem*; Pearson, *Black Scientists*; Ivan Van Sertima, *Blacks in Science: Ancient and Modern* (New Brunswick, N.J.: Transaction Books, 1986); *Science*. A recent *Los Angeles Times* article on the scientists and engineers responsible for the Mars space probe shows a racially diverse collection of young people (April 11, 2004).

22. See Hine, "Co-laborers in the Work of the Lord."

23. *Science.*

24. See Ashis Nandy, "Science as a Reason of State," in *Science, Hegemony, and Violence: A Requiem for Modernity*, ed. Nandy (New Delhi, India: Oxford University Press, 1990).

25. Michael Adas, *Machines as the Measure of Man* (Ithaca: Cornell University Press, 1989).

26. I discuss this issue in "Cultures as Toolboxes for Sciences and Technologies," chap. 4 of *Is Science Multicultural? Postcolonialisms, Feminisms, and Epistemologies* (Bloomington: Indiana University Press, 1998).

27. See Edward Said, *Orientalism* (New York: Pantheon, 1978).

28. I borrow the notion of a political unconscious from Fredric Jameson, *The Political Unconscious: Narrative as a Socially Symbolic Act* (Ithaca: Cornell University Press, 1981).

29. The phrase is Laura Nader's, in the introduction to her edited collection *Naked Science: Anthropological Inquiry into Boundaries, Power, and Knowledge* (New York: Routledge, 1996).

Chapter 2. Seeing Ourselves as Others See Us: Postcolonial Science Studies

1. These issues were explored in a somewhat different way in the first four chapters of my *Is Science Multicultural? Postcolonialisms, Feminisms, and Epistemologies* (Bloomington: Indiana University Press, 1998).

2. The proceedings of two international conferences, a manifesto by Third World science scholars, and an encyclopedia each provide a good introduction to this field: see Patrick Petitjean, Catherine Jami, and Anne Marie Moulin, eds., *Science and Empires: Historical Studies about Scientific Development and European Expansion* (Dordrecht: Kluwer, 1992); Ziauddin Sardar, ed., *The Revenge of Athena: Science, Exploitation, and the Third World* (London: Mansell, 1988); Third World Network, *Modern Science in Crisis: A Third World Response* (Penang, Malaysia: Third World Network, 1988), reprinted in Sandra Harding, ed., *The "Racial" Economy of Science: Toward a Democratic Future* (Bloomington: Indiana University Press, 1993); and Helaine Selin, ed., *Encyclopedia of the History of Science, Technology, and Medicine in Non-Western Cultures* (Dordrecht: Kluwer, 1997). See also Harding, *Is Science Multicultural?* Robert Figueroa and Sandra Harding, eds., *Science and Other Cultures: Issues in Philosophies of Science and Technology* (New York: Routledge, 2003); David J. Hess, *Science and Technology in a Multicultural World: The Cultural Politics of Facts and Artifacts* (New York: Columbia University Press, 1995); and Laura Nader, ed., *Naked Science: Anthropological Inquiry into Boundaries, Power, and Knowledge* (New York: Routledge, 1996).

3. See Selin, *Encyclopedia of Science, Technology, and Medicine*.

4. See Frances Yates, *Giordano Bruno and the Hermetic Tradition* (New York: Vintage, 1969); Donald F. Lach, *Asia in the Making of Europe*, vol. 2 (Chicago: University of Chicago Press, 1977); Seyyed Hossein Nasr, "Islamic Science, Western Science: Com-

mon Heritage, Diverse Destinies," in *Revenge of Athena*, ed. Sardar; Martin Bernal, *Black Athena: The Afroasiatic Roots of Classical Civilization*, vol. 1 (New Brunswick, N.J.: Rutgers University Press, 1987); J. M. Blaut, *The Colonizer's Model of the World: Geographical Diffusionism and Eurocentric History* (New York: Guilford Press, 1993); and I. A. Sabra, "The Scientific Enterprise," in *The World of Islam*, ed. B. Lewis (London: Thames and Hudson, 1976).

5. Jack Weatherford, *Indian Givers: What the Native Americans Gave to the World* (New York: Crown, 1988).

6. R. K. Kochhar, "Science in British India," pt. 1, *Current Science* (India) 63.11 (1992): 694. Cf. also pt. 2, 64.1 (1993): 55–62. I thank Joni Seager for directing me to this essay.

7. And, as V. Y. Mudimbe pointed out to me in conversation, the same processes occurred in Europe itself, for European sciences also constituted European lands, cities, and peoples as their laboratories. Consider, for example, the way women, the poor, children, the sick, the mad, rural and urban populations, and workers have been continuously studied by natural and social sciences.

8. Susantha Goonatilake makes this point in "The Voyages of Discovery and the Loss and Rediscovery of 'Other's' Knowledge," *Impact of Science on Society*, no. 167 (1992): 241–64. See also his *Aborted Discovery: Science and Creativity in the Third World* (London: Zed Press, 1984) and *Toward a Global Science: Mining Civilizational Knowledge* (Bloomington: Indiana University Press, 1998).

9. The British anthropologist Bronislaw Malinowski points out that the scientific impulse must have been present in every culture, intertwined, as in modern European cultures, with magic and religion. He proposes that "primitive humanity was aware of the scientific laws of natural process, [and] that all people operate within the domains of magic, science, and religion" (*Magic, Science, and Religion and Other Essays* [1925; reprint, Garden City, N.Y.: Doubleday Anchor, 1948], 196, as quoted in Nader, *Naked Science*, 5).

10. Nancy Brickhouse's questions helped me to clarify this point.

11. Susantha Goonatilake, "A Project for Our Times," in *Revenge of Athena*, ed. Sardar, 230.

12. See the section on astronomy in Ivan Van Sertima, *Blacks in Science: Ancient and Modern* (New Brunswick, N.J.: Transaction Books, 1986).

13. See, for example, George Gheverghese Joseph, *The Crest of the Peacock: Non-European Roots of Mathematics* (New York: I. B. Tauris, 1991).

14. Goonatilake, "Voyages of Discovery," 246.

15. Joseph Needham, *The Grand Titration: Science and Society in East and West* (Toronto: University of Toronto Press, 1969), 55–56. Blaut makes the same point in *Colonizer's Model*.

16. See, for instance, Susan Bordo, *The Flight to Objectivity* (Albany: SUNY Press, 1987); Genevieve Lloyd, *The Man of Reason: "Male" and "Female" in Western Philosophy* (Minneapolis: University of Minnesota Press, 1984); Tzvetan Todorov, *The Conquest of America: The Question of the Other*, trans. Richard Howard (New York: Harper

and Row, 1984); and Eric Wolf, *Europe and the Peoples without a History* (Berkeley and Los Angeles: University of California Press, 1984).

17. David Hess also makes this point in the introduction to his *Science and Technology*. See Needham's discussion (in *Grand Titration*) of seven conceptual errors in standard Western thought about "universal science" that lead to erroneous devaluations of the scientific achievements of non-European sciences.

18. Robin Horton, "African Traditional Thought and Western Science," pts. 1 and 2, *Africa* 37 (1967): 50–71, 155–87; J. E. Wiredu, "How Not to Compare African Thought with Western Thought," in *African Philosophy*, ed. Richard Wright, 3d ed. (Lanham, Md.: University Press of America, 1984). See Helen Verran's recent challenge to the view that the separation of words from the world (from how the world is initially "learned") is a distinctive element of science in her *Science and an African Logic* (Chicago: University of Chicago Press, 2001). For religious elements of Western science, see David Noble's work, for example, *A World without Women: The Christian Clerical Culture of Western Science* (New York: Alfred A. Knopf, 1992) and *The Religion of Technology* (New York: Alfred A. Knopf, 1995).

19. Thomas S. Kuhn, *The Structure of Scientific Revolutions*, 2d ed. (Chicago: University of Chicago Press, 1970), 167.

20. Needham, *Grand Titration*, 14–15.

21. Edgar Zilsel, "The Sociological Roots of Science," *American Journal of Sociology* 47 (1942).

22. See Edwin Hutchins, *Cognition in the Wild* (Cambridge: MIT Press, 1995).

23. For such accounts, some earlier than the quincentennial, see Blaut, *Colonizer's Model;* Lucille H. Brockway, *Science and Colonial Expansion: The Role of the British Royal Botanical Gardens* (New York: Academic Press, 1979); Hess, *Science and Technology;* James E. McClellan, *Colonialism and Science: Saint Domingue in the Old Regime* (Baltimore: Johns Hopkins University Press, 1992); Petitjean, Jami, and Moulin, *Science and Empires;* and Weatherford, *Indian Givers.*

24. Kochhar, "Science in British India"; Alfred Crosby, *Ecological Imperialism: The Biological Expansion of Europe* (Westport, Conn.: Greenwood Press, 1972); V. V. Krishna, "The Colonial 'Model' and the Emergence of National Science in India, 1876–1920," in *Science and Empires*, ed. Petitjean, Jami, and Moulin; Deepak Kumar, "Problems in Science Administration: A Study of the Scientific Surveys in British India, 1757–1900" (Dordrecht: Kluwer, 1992).

25. Paul Forman, "Behind Quantum Electronics: National Security as Bases for Physical Research in the U.S., 1940–1960," *Historical Studies in Physical and Biological Sciences* 18 (1987): 149–229; Wolfgang Sachs, ed., *The Development Dictionary: A Guide to Knowledge as Power* (Atlantic Highlands, N.J.: Zed Press, 1992).

26. Goonatilake, "A Project for Our Times," 235–36. (Shouldn't African and indigenous American civilizations also count as regional ones containing scientific traditions?)

27. The "strong programme" in the sociology of knowledge developed this kind

of analysis. See, for example, David Bloor, *Knowledge and Social Imagery* (London: Routledge and Kegan Paul, 1977). It also unnecessarily embraced relativism, an issue pursued further in chapter 9. Analyses of the complex social and natural origins of scientific claims that also argue against relativism have emerged from many theorists. See, for example, Ulrich Beck, *Risk Society: Towards a New Modernity* (London: Sage, 1992); Donna Haraway, "Situated Knowledges: The Science Question in Feminism and the Privilege of Partial Perspectives," in her *Simians, Cyborgs, and Women: The Reinvention of Nature* (New York: Routledge, 1991); and Bruno Latour, *We Have Never Been Modern,* trans. Catherine Porter (Cambridge: Harvard University Press, 1993).

28. I have written on the topic of this section in many places. An earlier version of the arguments here appeared in "Is Modern Science an Ethnoscience?" in *Sociology of the Sciences Yearbook,* ed. T. Shinn, J. Spaapen, and Raoul Waast (Dordrecht: Kluwer, 1996); and "Cultures as Toolboxes," chap. 4 of *Is Science Multicultural?* 57–61.

29. For an illuminating account of the environmental problems European expansion solved for feudal Europe, see Jason W. Moore, "The Crisis of Feudalism: An Environmental History," *Organization and Environment* 15.3 (2002): 301–22. Unfortunately, Moore points out, in the world today we can look forward to no such convenient resolution to the reappearance of similar problems.

30. See Bruno Latour's discussion of the importance to science of "centres of calculation," in chap. 6 of his *Science in Action* (Cambridge: Harvard University Press, 1987). For extensive study of one such center, Kew Gardens, see Brockway, *Science and Colonial Expansion.*

31. See, for example, Morris Berman, *The Reenchantment of the World* (Ithaca: Cornell University Press, 1981); Bordo, *The Flight to Objectivity;* Carolyn Merchant, *The Death of Nature: Women, Ecology, and the Scientific Revolution* (New York: Harper and Row, 1980); and Nasr, "Islamic Science."

32. Needham, *Grand Titration,* 302, 323, 327.

33. Evelyn Fox Keller, *Reflections on Gender and Science* (New Haven: Yale University Press, 1984), 131, 132. See also the interesting discussion of Needham's argument in Jatinder K. Bajaj, "Francis Bacon, the First Philosopher of Modern Science: A Non-Western View," in *Science, Hegemony, and Violence: A Requiem for Modernity,* ed. Ashis Nandy (New Delhi, India: Oxford University Press, 1990).

34. Kok Peng Khor, "Science and Development: Underdeveloping the Third World," in *Revenge of Athena,* ed. Sardar, 207–8.

35. Claude Alvares, "Science, Colonialism, and Violence: A Luddite View," in *Science, Hegemony, and Violence,* ed. Nandy, 108.

36. J. Bandyopadhyay and Vandana Shiva, "Science and Control: Natural Resources and Their Exploitation," in *Revenge of Athena,* ed. Sardar, 63.

37. Dorothy Smith is especially eloquent on this point. See *The Conceptual Practices of Power: A Feminist Sociology of Knowledge* (Boston: Northeastern University Press, 1990) and *The Everyday World as Problematic: A Sociology for Women* (Boston: Northeastern University Press, 1987). However, abstractness is not unique to mod-

ern Western cultures. As Paola Bachetta pointed out (by letter), certain forms of ancient Hinduism are based on philosophical abstractions.

38. Bandyopadhyay and Shiva, "Science and Control," 60.

39. Xavier Polanco, "World-Science: How Is the History of World-Science to Be Written?" in *Science and Empires,* ed. Petitjean, Jami, and Moulin, 225.

40. Goonatilake, "A Project for Our Times," 229–30.

Chapter 3. With Both Eyes Open: A World of Sciences

1. Bronislaw Malinowski, *Magic, Science, and Religion, and Other Essays* (1925; reprint, Garden City, N.Y.: Doubleday Anchor, 1948).

2. See Peter Galison and David J. Stump, eds., *The Disunity of Science: Boundaries, Contexts, and Power* (Stanford: Stanford University Press, 1996).

3. Edward O. Wilson, *Consilience: The Unity of Knowledge* (New York: Alfred A. Knopf, 1998), 5.

4. See, for example, Meera Nanda, *Prophets Facing Backward: Postmodern Critiques of Science and Hindu Nationalism in India* (New Brunswick, N.J.: Rutgers University Press, 2004).

5. Laura Nader recommends "both eyes open" in "Anthropological Inquiry into Boundaries, Power, and Knowledge," the introduction to *Naked Science: Anthropological Inquiry into Boundaries, Power, and Knowledge,* ed. Nader (New York: Routledge, 1996).

6. I discuss this issue in "Cultures as Toolboxes," chap. 4 of *Is Science Multicultural? Postcolonialisms, Feminisms, and Epistemologies* (Bloomington: Indiana University Press, 1998).

7. See, for example, Uma Narayan, "The Project of a Feminist Epistemology: Perspectives from a Nonwestern Feminist," in *Gender/Body/Knowledge,* ed. Susan Bordo and Alison Jaggar (New Brunswick, N.J.: Rutgers University Press, 1989).

8. Susantha Goonatilake, "The Voyages of Discovery and the Loss and Rediscovery of the 'Other's' Knowledge," *Impact of Science on Society,* no. 167 (1992): 241–64. See also his *Toward a Global Science: Mining Civilizational Knowledge* (Bloomington: Indiana University Press, 1998).

9. This language has been developed by Samir Amin in his *Eurocentrism* (New York: Monthly Review, 1989).

10. Third World Network, *Modern Science in Crisis: A Third World Response* (Penang, Malaysia: Third World Network, 1988), reprinted in Sandra Harding, ed., *The "Racial" Economy of Science: Toward a Democratic Future* (Bloomington: Indiana University Press, 1993), 487, 495.

11. Elizabeth Hsu, "The Reception of Western Medicine in China: Examples from Yunnan," in *Science and Empires: Historical Studies about Scientific Development and European Expansion,* ed. Patrick Petitjean, Catherine Jami, and Anne Marie Moulin (Dordrecht: Kluwer, 1992). See also Andrew Feenberg, "Technology in a Global

World"; and Junichi Murata, "Creativity of Technology and the Modernization Process of Japan," both in *Science and Other Cultures: Issues in Philosophies of Science and Technology*, ed. Robert Figueroa and Sandra Harding (New York: Routledge, 2003).

12. Ashis Nandy, ed., introduction to *Science, Hegemony, and Violence: A Requiem for Modernity* (New Delhi, India: Oxford University Press, 1990), 11.

13. The journal can be found at http://www.nuffic.ni/ciran/ikdm.html.

14. *Vital Signs: UCLA Healthcare*, no. 29 (summer 2003): 5. One could debate whether this is an example of adding non-Western to Western science or vice versa.

15. Ontario Museum of Science and Technology, *A Question of Truth*, museum catalog (Toronto: Ontario Science Centre, 1997). And see Helaine Selin, ed., *Encyclopedia of the History of Science, Technology, and Medicine in Non-Western Cultures* (Dordrecht: Kluwer, 1997).

16. Third World Network, *Modern Science in Crisis*, 31.

17. Richard Levins and Richard Lewontin, "Applied Biology in the Third World," in *The Dialectical Biologist* (Cambridge: Harvard University Press, 1988), reprinted in *"Racial" Economy of Science*, ed. Harding, 315–25.

18. See, for example, Ulrich Beck, *Risk Society: Towards a New Modernity* (London: Sage, 1992); Bruno Latour, *We Have Never Been Modern*, trans. Catherine Porter (Cambridge: Harvard University Press, 1993); and Nandy, *Science, Hegemony, and Violence*.

19. See, for example, C. P. Snow, *The Two Cultures: And a Second Look* (1959; reprint, Cambridge: Cambridge University Press, 1964). The following argument draws from my "Women and Science in Historical Context," *NWSA Journal* 5.1 (1993): 49–55.

20. Jerome Ravetz had argued for such a similar set of courses in his *Scientific Knowledge and Its Social Problems* (New York: Oxford University Press, 1971).

21. Beck, in *Risk Society*, has made a similar argument for the development of a "second, reflexive modernity" to complete the tasks that the "first, industrial modernity" has been fated to leave undone.

22. Fredric Jameson makes a similar point about the value of the "principled relativism" of feminist standpoint theory in his *"History and Class Consciousness* as an 'Unfinished Project,'"* in *The Feminist Standpoint Theory Reader: Intellectual and Political Controversies*, ed. Sandra Harding (New York: Routledge, 2003).

Chapter 4. Northern Feminist Science Studies: New Challenges and Opportunities

1. See Uma Narayan, "The Project of a Feminist Epistemology: Perspectives from a Nonwestern Feminist," in *Gender/Body/Knowledge*, ed. Susan Bordo and Alison Jaggar (New Brunswick, N.J.: Rutgers University Press, 1989); and Meera Nanda, *Prophets Facing Backward: Postmodern Critiques of Science and Hindu Nationalism in India* (New Brunswick, N.J.: Rutgers University Press, 2004).

2. See Sandra Harding, "Gender and Science: Two Problematic Concepts," in *The Science Question in Feminism* (Ithaca: Cornell University Press, 1986), for an early for-

mulation of the importance of thinking of gender as an attribute of social structures and meaning systems, as well as of individual men and women. Unfortunately, the tendency to use the term to refer only to individuals, and even only to women, has weakened many attempts to advance gender justice. One notable example is the directive to all United Nations agencies to conduct gender audits of their programs. These are almost invariably understood as audits of the representation of women in the administration and client pool of the programs. To be sure, increasing the representation of women in these ways can be a valuable project. But it avoids dealing with the harder issues of men's perception around the world that male supremacy is their entitlement by birth, and how such male-supremacy assumptions shape institutional practices, meanings, and policies. See Robert W. Connell's analysis, "Change among the Gatekeepers: Men, Masculinities, and Gender Equality in the Global Arena," *Signs: Journal of Women in Culture and Society* 30.3 (2005): 1801–26.

3. See Margaret Rossiter, *Women Scientists in America: Struggles and Strategies to 1940* (Baltimore: Johns Hopkins University Press, 1982); Londa Schiebinger, *The Mind Has No Sex? Women in the Origins of Modern Science* (Cambridge: Harvard University Press, 1989); Schiebinger, *Nature's Body: Gender in the Making of Modern Science* (Boston: Beacon Press, 1993); and Rossiter, *Women Scientists in America: Before Affirmative Action* (Baltimore: Johns Hopkins University Press, 1995).

4. For a recent overview of the U.S. programs, see American Association of University Women Educational Foundation, *Under the Microscope: A Decade of Gender Equity Projects in the Sciences* (Washington, D.C.: AAUW Educational Foundation, 2004); and National Science Foundation, *New Formulas for America's Workforce: Girls in Science and Engineering* (Washington, D.C.: National Science Foundation Education and Human Resources Directorate, 2004). For some of the European Union's work, see the proceedings of a European Commission conference in Brussels, November 8–9, 2001, *Gender and Research,* ed. Linda Maxwell, Karen Slavin, and Kerry Young (Brussels: European Commission, 2002); and European Commission, *Waste of Talents: Turning Private Struggles into a Public Issue: Women and Science in the ENWISE Countries* (Luxembourg: European Communities, 2003).

5. *A Study on the Status of Women Faculty in Science at MIT* (Cambridge: MIT, 1999), available online at http://web.mit.edu/fnl/women/woen.html. See American Association of University Women Educational Foundation, *Under the Microscope,* for an overview and critique of these studies. See also Henry Etzkowitz, Carol Kemelgor, and Brian Uzzi, *Athena Unbound: The Advancement of Women in Science and Technology* (Cambridge: Cambridge University Press, 2000); and Sue V. Rosser, *Women, Science, and Society: The Crucial Union* (New York: Teacher's College Press, 2000).

6. In highly gender-segregated cultures, such as some Latin American and Islamic ones, aristocratic women sometimes gain educations and credentials more like their brothers than is the case in less gender-segregated cultures. Additionally, women must become physicians if girls and women are to receive medical attention in such cultures.

7. See Sandra Harding and Elizabeth McGregor, "The Gender Dimension of Sci-

ence and Technology," in *UNESCO World Science Report,* ed. Howard J. Moore (Paris: UNESCO, 1996). The Third World Organization for Women in Science provides an important international network "supporting the advancement of girls and women in science and technology." It conveys news of jobs and sponsors national chapters, fellowships, workshops, conferences, online dialogues, and so on. Electronic versions of its newsletter may be found at http://www.ictp.trieste.it/twas/twows.html.

8. See Kathleen Broome Williams, *Improbable Warriors: Women Scientists and the U.S. Navy in World War II* (Annapolis, Md.: Naval Institute Press, 2001); Margaret A. M. Murray, *Women Becoming Mathematicians: Creating a Professional Identity in Post–World War II America* (Cambridge: MIT Press, 2000); Barbara R. Stein, *On Her Own Terms: Annie Montague Alexander and the Rise of Science in the American West* (Berkeley and Los Angeles: University of California Press, 2001); and Brenda Maddox, *Rosalind Franklin: The Dark Lady of DNA* (New York: HarperCollins, 2002).

9. Ruth Hubbard, "Science, Power, Gender: How DNA Became the Book of Life," *Signs: Journal of Women in Culture and Society* 28.3 (2003): 798. References are to Anne Sayre, *Rosalind Franklin and DNA* (New York: W. W. Norton, 1975); and James D. Watson, *The Double Helix* (New York: Atheneum, 1968).

10. Hubbard, "Science, Power, Gender," 798.

11. In *Signs: Journal of Women in Culture and Society* 28:3 (2003). Hammonds's next book is *The Logic of Difference: A History of Race in Science and Medicine in the United States.* Subramaniam is working on a book called *A Question of Variation: Race, Gender and the Practice of Science,* focused on how experimental biologists can engage the practices of feminist science studies.

12. Environmental research and AIDS research are just two recent examples of the many projects that have had a similar start-up pattern.

13. Ethel Tobach and Betty Rosoff, eds., *Genes and Gender,* vols. 1–4 (New York: Gordian Press, 1978–1984).

14. See Anne Fausto-Sterling, "The Myth of Neutrality: Race, Sex, and Class in Science," *Radical Teacher* 19 (1981): 21–25; "Race, Gender, and Science," *Transformations* 2.2 (1991): 4–12; and "Gender, Race, and Nation: The Comparative Anatomy of Hottentot Women in Europe I: 1815–1817," in *Deviant Bodies,* ed. Jennifer Terry and Jacqueline Urla (Bloomington: Indiana University Press, 1995). When I first taught a course focused on antiracist as well as antisexist issues in the sciences in the early 1980s, Fausto-Sterling's syllabus for her course, which she kindly shared, was the only one any feminist scholar of my acquaintance had heard of. She and Harvard historian of science Evelynn Hammonds, especially, have waged a continuing struggle since then to get such issues addressed in feminist science studies contexts as well as in biological, health, and public health sciences themselves. See also my edited collection *The "Racial" Economy of Science: Toward a Democratic Future* (Bloomington: Indiana University Press, 1993) and my *Is Science Multicultural? Postcolonialisms, Feminisms, and Epistemologies* (Bloomington: Indiana University Press, 1998).

15. Anne Fausto-Sterling, "Refashioning Race: DNA and the Politics of Health

Care," *Differences: A Journal of Feminist Cultural Studies* 15.3 (1994): 1–37. See also her essay "The Bare Bones of Sex: Part I—Sex and Gender," *Signs: Journal of Women in Culture and Society* 30.2 (2005): 1491–1528.

16. See Donna Haraway, *Primate Visions: Gender, Race, and Nature in the World of Modern Science* (New York: Routledge, 1989); *Simians, Cyborgs, and Women: The Reinvention of Nature* (New York: Routledge, 1991); *Modest_Witness@Second_Millennium.FemaleMan_Meets_OncoMouse: Feminism and Technoscience* (New York: Routledge, 1997); and *The Companion Species Manifesto: Dogs, People, and Significant Otherness* (Chicago: Prickly Paradigm Press, 2003), 3.

17. Anna Wilson, "Sexing the Hyena: Intraspecies Readings of the Female Phallus," *Signs: Journal of Women in Culture and Society* 28.3 (2003): 755–90; Elizabeth D. Harvey, "Anatomies of Rapture: Clitoral Politics/Medical Blazons," *Signs: Journal of Women in Culture and Society* 27.2 (2002): 315–46; Nancy Tuana, "Coming to Understand: Orgasm and the Epistemology of Ignorance," *Hypatia: A Journal of Feminist Philosophy* 19.1 (2004): 194–232; Rachel P. Maines, *The Technology of Orgasm: "Hysteria," the Vibrator, and Women's Sexual Satisfaction* (Baltimore: Johns Hopkins University Press, 1999).

18. Joni Seager, "Rachel Carson Died of Breast Cancer: The Coming of Age of Feminist Environmentalism," *Signs: Journal of Women in Culture and Society* 28.3 (2003): 945–72. See also Seager's earlier book, *Earth Follies: Coming to Feminist Terms with the Global Environmental Crisis* (New York: Routledge, 1993); and Rosi Braidotti, E. Charkiewicz, Sabina Hausler, and Saskia Wieringa, *Women, the Environment, and Sustainable Development* (Atlantic Highlands, N.J.: Zed Press, 1994), which also contains an account of the strengths and limitations of ecofeminism, and of the limitations of mainstream environmental movements with respect to feminist concerns, to which we return in the next chapter.

19. Peter Singer, *Animal Liberation* (New York: Avon, 1975); Tom Regan, *The Case for Animal Rights* (Berkeley and Los Angeles: University of California Press, 1983); Seager, "Rachel Carson," 953, 951 (Haraway's studies also expose these perceptions and assumptions); Vandana Shiva, *Stolen Harvest: The Hijacking of the Global Food Supply* (Cambridge, Mass.: South End Press, 2000), 72–75. Shiva's *Staying Alive: Women, Ecology, and Development* (London: Zed Press, 1989) played an important role in bringing Third World women's perspectives on the environment and on development policies into women's studies classrooms.

20. Vandana Shiva, *Close to Home: Women Reconnect Ecology, Health, and Development Worldwide* (Philadelphia: New Society, 1994), 3, quoted in Seager, "Rachel Carson," 959.

21. Seager, "Rachel Carson," 964. Seager recommends the articulations of the principle in Sandra Steingraber, *Living Downstream* (New York: Vintage, 1997); Carolyn Raffensperger and Joel Tickner, eds., *Protecting Public Health and the Environment: Implementing the Precautionary Principle* (Washington, D.C.: Island Press, 1999); and the Web site of the Science and Environmental Health Network, http://www.sehn.org.

22. Seager, "Rachel Carson," 967.

23. Haraway, *Primate Visions;* Alison Wylie, "The Constitution of Archaeological Evidence: Gender Politics and Science" (Stanford: Stanford University Press, 1996); Wylie, *Gender, Politics, and Scientific Archaeology: The Challenge of "Gender Research" in Archaeology* (New York: Blackwell, 2004); Evelyn Fox Keller, *A Feeling for the Organism* (San Francisco: Freeman, 1983); Londa Schiebinger, Angela N. H. Creager, and Elizabeth Lunbeck, eds., *Feminism in Twentieth-Century Science, Technology, and Medicine* (Chicago: University of Chicago Press, 1991). See also three essays Schiebinger brought together in *Signs: Journal of Women in Culture and Society* 28.3 (2003): Margaret W. Conkey, "Has Feminism Changed Archaeology?" (867–80); Amy Bug, "Has Feminism Changed Physics?" (881–900); and Patricia Adair Gowaty, "Sexual Natures: How Feminism Changed Evolutionary Biology" (901–22).

24. Sharon Traweek has also written about this phenomenon among Japanese women scientists.

25. Maralee Mayberry, Banu Subramaniam, and Lisa H. Weasel, eds., *Feminist Science Studies: A New Generation* (New York: Routledge, 2001); Mary Wyer, Mary Barbercheck, Donna Giesman, Hatice Örün Öztürk, and Marta Wayne, *Women, Science, and Technology: A Reader in Feminist Science Studies* (New York: Routledge, 2001).

26. It is beyond the scope of this chapter to review this rich field. For one early sample of issues, see Sandra Harding, ed., *Feminism and Methodology: Social Science Issues* (Bloomington: Indiana University Press, 1987). See also *New Feminist Approaches to Social Science Methodology,* a special issue of *Signs: Journal of Women in Culture and Society* 30.4 (2005), ed. Sandra Harding and Kate Norberg.

27. Carolyn Merchant, *The Death of Nature: Women, Ecology, and the Scientific Revolution* (New York: Harper and Row, 1980); Schiebinger, *Mind Has No Sex?* Schiebinger, *Nature's Body;* Elizabeth Potter, *Gender and Boyle's Law of Gases* (Bloomington: Indiana University Press, 2001); Evelyn Fox Keller, *Reflections on Gender and Science* (New Haven: Yale University Press, 1984); Sharon Traweek, *Beamtimes and Life Times* (Cambridge: MIT Press, 1988); Bonnie Spanier, *Im/partial Science: Gender Ideology in Molecular Biology* (Bloomington: Indiana University Press, 1995).

28. American Association of University Women Educational Foundation, *Under the Microscope;* Nancy Brickhouse, "Bringing in the Outsiders: Reshaping the Sciences of the Future," *Journal of Curriculum Studies* 26.4 (1994): 401–16; Alison Kelly, ed., *The Missing Half: Girls and Science Education* (Manchester: Manchester University Press, 1981); Kelly, ed., *Science for Girls?* (Philadelphia: Open University Press, 1987); Sue V. Rosser, "Female Friendly Science: Including Women in Curricular Content and Pedagogy in Science," *Journal of General Education* 42.3 (1993): 191–220; Rosser, *Teaching Science and Health from a Feminist Perspective: A Practical Guide* (New York: Pergamon, 1986).

29. Nancy Brickhouse, "Embodying Science: A Feminist Perspective on Learning" *Journal of Research in Science Teaching* 38.3 (2001): 282–95.

30. Nancy Brickhouse reports that "girls are as likely to be enrolled in some ad-

vanced high school science courses as boys (AAUW 1998) and are as numerous in some scientific college majors as boys (e.g., biology). Furthermore, sex differences in achievement are small or nonexistent (Third International Mathematics and Science Study 1997)" (ibid., 282). See American Association of University Women Educational Foundation, *Gender Gap: Where Schools Still Fail Our Children* (New York: Marlowe, 1998), and their recent report on gender-equity projects in the sciences, *Under the Microscope.*

31. Nancy Brickhouse, "Feminism(s) and Science Education," *International Handbook of Science Education* (1998): 1067–81, provides a good review of an array of such issues.

32. Cynthia Cockburn, *Machinery of Dominance: Women, Men, and Technical Know-How* (London: Pluto Press, 1985); Judy Wajcman, *Feminism Confronts Technology* (University Park: Pennsylvania State University Press, 1991). See also Adele E. Clarke and Virginia L. Olesen, eds., *Revisioning Women, Health, and Healing: Feminist, Cultural, and Technoscience Perspectives* (New York: Routledge, 1999); Janine Marchessault and Kim Sawchuk's collection of studies of visual and imaging technologies, *Wild Science: Reading Feminism, Medicine, and the Media* (London: Routledge, 2000); and David Noble, *The Religion of Technology* (New York: Alfred A. Knopf, 1995).

33. In different ways this was an insight of both Foucault and Kuhn. It is developed in Ian Hacking, *Representing and Intervening* (Cambridge: Cambridge University Press, 1983); Joseph Rouse, *Knowledge and Power: Toward a Political Philosophy of Science* (Ithaca: Cornell University Press, 1987); Rouse, *Engaging Science: How to Understand Its Practices Philosophically* (Ithaca: Cornell University Press, 1996); and Rouse, *How Scientific Practices Matter: Reclaiming Philosophical Naturalism* (Chicago: University of Chicago Press, 2002).

Chapter 5. Discriminatory Epistemologies and Philosophies of Science

1. Fredric Jameson examined the "political unconscious" of the novel in *The Political Unconscious: Narrative as a Socially Symbolic Act* (Ithaca: Cornell University Press, 1981).

2. Note that it is masculinity also, not just femininity, that is the issue in these accounts. Recollect that environmental theorists Peter Singer and Tom Regan distanced their form of environmentalism from forms associated with womanly "caring," as Joni Seager reported in "Rachel Carson Died of Breast Cancer: The Coming of Age of Feminist Environmentalism," *Signs: Journal of Women in Culture and Society* 28.3 (2003): 945–72. See David Noble, *A World without Women: The Christian Clerical Culture of Western Science* (New York: Alfred A. Knopf, 1992) and *The Religion of Technology* (New York: Alfred A. Knopf, 1995), for two influential historical studies of the masculinity of modern sciences and technology.

3. However, as we shall see, Sharyn Clough recommends abandoning such gen-

eral epistemological projects in her *Beyond Epistemology: A Pragmatist Approach to Feminist Science Studies* (Lanham, Md.: Rowman and Littlefield, 2003).

4. Dorothy Smith's earliest essay was published in 1974. This and other early and later ones are collected in *The Everyday World as Problematic: A Sociology for Women* (Boston: Northeastern University Press, 1987); *The Conceptual Practices of Power: A Feminist Sociology of Knowledge* (Boston: Northeastern University Press, 1990); *Texts, Facts, and Femininity: Exploring the Relations of Ruling* (New York: Routledge, 1990); and *Writing the Social: Critique, Theory, and Investigations* (Toronto: University of Toronto Press, 1999). See also Nancy Hartsock, "The Feminist Standpoint: Developing the Ground for a Specifically Feminist Historical Materialism," in *Discovering Reality: Feminist Perspectives on Epistemology, Metaphysics, Methodology, and Philosophy of Science*, ed. Sandra Harding and Merrill Hintikka (Dordrecht: Reidel, 1983); and Hilary Rose, "Hand, Brain, and Heart: A Feminist Epistemology for the Natural Sciences," *Signs: Journal of Women in Culture and Society* 9.1 (1983): 73–90.

5. Central writings here are Alison Jaggar, "Feminist Politics and Epistemology: Justifying Feminist Theory," chap. 11 of *Feminist Politics and Human Nature* (Totowa, N.J.: Rowman and Allenheld, 1983); Patricia Hill Collins, *Black Feminist Thought: Knowledge, Consciousness, and the Politics of Empowerment* (New York: Routledge, 1991); Sandra Harding, *The Science Question in Feminism* (Ithaca: Cornell University Press, 1986); Donna Haraway, "Situated Knowledges," in *Simians, Cyborgs, and Women: The Reinvention of Nature* (New York: Routledge, 1991) (this was first presented at an American Philosophical Association meeting in 1987 as a comment on my *Science Question*); and Harding, "Rethinking Standpoint Epistemology: What Is 'Strong Objectivity'?" in *Feminist Epistemologies*, ed. Linda Alcoff and Elizabeth Potter (New York: Routledge, 1992). See also Harding, ed., *The Feminist Standpoint Theory Reader: Intellectual and Political Controversies* (New York: Routledge, 2003).

6. However, as Dorothy Smith has pointed out, my writings about standpoint theory as a general type of epistemology, methodology, and philosophy of science have tended to obscure important disciplinary differences in these theorists' projects ("Comment on Hekman's 'Truth and Method: Feminist Standpoint Theory Revisited,'" *Signs: Journal of Women in Culture and Society* 22.2 [1997]: 392–98).

7. See Haraway, "Situated Knowledges."

8. See Smith, *Conceptual Practices.*

9. Hartsock, "Feminist Standpoint," 288.

10. The phrase "strong objectivity" is mine; see "Rethinking Standpoint Epistemology: What Is 'Strong Objectivity'?" But other feminists have also refashioned the notion in distinctive ways.

11. Most of these debates can be found in Harding, *Feminist Standpoint Theory Reader.*

12. Alison Wylie, "Why Standpoint Matters," in *Science and Other Cultures: Issues in Philosophies of Science and Technology*, ed. Robert Figueroa and Sandra Harding (New York: Routledge, 2003), 339–40.

13. See, for example, many of the essays in Harding, *Feminist Standpoint Theory Reader*, especially Fredric Jameson, "*History and Class Consciousness* as an 'Unfinished Project'"; Dick Pels, "Strange Standpoints; or, How to Define the Situation for Situated Knowledge"; Nancy J. Hirschmann, "Feminist Standpoint as Postmodern Strategy"; and Joseph Rouse, "Feminism and the Social Construction of Scientific Knowledge."

14. Dorothy Smith has argued for the importance of understanding that her work is engaged in ongoing discussions in sociology, and that the very concept of diverse kinds of "standpoint theory" is an artifact of my general arguments about similarities between her work and that of others. See note 6.

15. Joseph Rouse makes this argument in a different respect in his "Feminism and Social Construction," in *Feminist Standpoint Theory Reader*, ed. Harding.

16. See, for example, Helen Longino, *Science as Social Knowledge* (Princeton: Princeton University Press, 1990).

17. The relevant excerpt from Jameson's "*History and Class Consciousness* as an 'Unfinished Project'" appears in *Feminist Standpoint Theory Reader*, ed. Harding, 144. His citation here is to Hartsock, "Feminist Standpoint"; Harding, *Science Question;* and Jaggar, "Feminist Politics and Epistemology." Hartsock's "Feminist Standpoint" also provides a clear account of standpoint theory's Marxian origins.

18. All of these are discussed in Harding, *Feminist Standpoint Theory Reader.*

19. See Mario Biagioli, ed., *The Science Studies Reader* (New York: Routledge, 1999).

20. See Joseph Rouse, *Knowledge and Power: Toward a Political Philosophy of Science* (Ithaca: Cornell University Press, 1987); *Engaging Science: How to Understand Its Practices Philosophically* (Ithaca: Cornell University Press, 1996); *How Scientific Practices Matter: Reclaiming Philosophical Naturalism* (Chicago: University of Chicago Press, 2002); "Feminism and the Social Construction of Scientific Knowledge," in *Feminism, Science, and the Philosophy of Science,* ed. Jack Nelson and Lynn Hankinson Nelson (Dordrecht: Kluwer, 1996), reprinted in Harding, *Feminist Standpoint Theory Reader;* and "Barad's Feminist Naturalism," *Hypatia: A Journal of Feminist Philosophy* 19.1 (2004): 142–61.

21. See, for example, Helaine Selin, ed., *Encyclopedia of the History of Science, Technology, and Medicine in Non-Western Cultures* (Dordrecht: Kluwer, 1997).

22. See Harding, "Robust Reflexivity," chap. 11 of *Is Science Multicultural? Postcolonialisms, Feminisms, and Epistemologies* (Bloomington: Indiana University Press, 1998).

23. Rouse, *Knowledge and Power,* 226–35.

24. See Karen Barad, "Meeting the Universe Halfway: Realism and Social Constructivism without Contradiction," in *Feminism, Science, and Philosophy of Science,* ed. Nelson and Nelson; "Getting Real: Technoscientific Practices and the Materialization of Reality," *Differences* 10.2 (1998): 87–128; "Agential Realism: Feminist Interventions in Understanding Scientific Practices," in *The Science Studies Reader,* ed. Biagioli; "Posthumanist Performativity: Toward an Understanding of How Matter

Comes to Matter," *Signs: Journal of Women in Culture and Society* 28.3 (2003): 801–32; and *Meeting the Universe Halfway* (forthcoming).

25. Clough, *Beyond Epistemology*, 113, 126. However, I fail to see how a "web of beliefs" escapes criticisms of representationalism. Isn't such a web a set of interlinked representations in both Quine's thought and Clough's? Clough focuses her criticisms on feminist science studies, and in particular on the early work (from the 1980s and early 1990s) of Helen Longino, Evelyn Fox Keller, and myself. I think she misreads this work and its projects in a variety of ways. For example, contrary to Clough's claim that feminist epistemology has been motivated by the desire to avoid skepticism, one of the most significant contrasts between feminist and mainstream epistemology is that the former was never motivated by such a desire, as Andrea Nye, for one, points out in *Philosophy and Feminism at the Border* (New York: Twayne, 1995), 85ff. And standpoint theory is not the "finding women's voices" project she presumes. Yet she certainly is right to spot problematic representationalist and epistemological projects in this early work.

26. See Rouse, "Feminism and Social Construction," in *Feminist Standpoint Theory Reader*, ed. Harding.

Chapter 6. Feminist Science and Technology Studies at the Periphery of the Enlightenment

1. See Gender Working Group, U.N. Commission on Science and Technology for Development, ed., *Missing Links: Gender Equity in Science and Technology for Development* (Ottawa: International Development Research Centre, 1995); Rosi Braidotti, E. Charkiewicz, Sabina Hausler, and Saskia Wieringa, *Women, the Environment, and Sustainable Development* (Atlantic Highlands, N.J.: Zed Press, 1994); Sandra Harding and Elizabeth McGregor, "The Gender Dimension of Science and Technology," in *UNESCO World Science Report*, ed. Howard J. Moore (Paris: UNESCO, 1996); and Vandana Shiva, *Staying Alive: Women, Ecology, and Development* (London: Zed Press, 1989).

2. Uma Narayan, "The Project of a Feminist Epistemology: Perspectives from a Nonwestern Feminist," in *Gender/Body/Knowledge*, ed. Susan Bordo and Alison Jaggar (New Brunswick, N.J.: Rutgers University Press, 1989).

3. Gender and feminist issues raised by these science and technology movements all tend to appear in every article and book on women's science and technology issues in the Global South. Important sources with a significant focus on the comparative ethnoscience issues are Braidotti et al., *Women and Sustainable Development*; Frederique Apffel-Marglin and Stephen A. Marglin, *Dominating Knowledge: Development, Culture, and Resistance* (Oxford: Clarendon Press, 1990); Shiva, *Staying Alive*; Shiva, *Monocultures of the Mind: Perspectives on Biodiversity and Biotechnology* (New York and Penang, Malaysia: Zed Press and Third World Network, 1993); and the journal *Gender, Technology, and Development* (Bangkok).

4. See Braidotti et al., *Women and Sustainable Development*; Maria Mies, *Patriarchy and Accumulation on a World Scale: Women in the International Division of Labor* (At-

lantic Highlands, N.J.: Zed Press, 1986); Shiva, *Staying Alive;* and Pamela Sparr, ed., *Mortgaging Women's Lives: Feminist Critiques of Structural Adjustment* (London: Zed Press, 1994). For recent analyses of women's situation in globalization and development, see Amrita Basu, Inderpal Grewal, Caren Kaplan, and Lisa Malkki, eds., *Globalization and Gender,* a special issue of *Signs: Journal of Women in Culture and Society* 26.4 (2001); and Francoise Lionnet, Obioma Nnaemeka, Susan Perry, and Celeste Schenk, eds., *Development Cultures: New Environments, New Realities, New Strategies,* a special issue of *Signs: Journal of Women in Culture and Society* 29.2 (2004).

5. Braidotti et al., *Women and Sustainable Development,* chaps. 5 and 7, gives a good sense of how the context of development agencies shaped the GESD discussion.

6. Drucilla K. Barker, "Dualisms, Discourse, and Development," in *Decentering the Center: Philosophy for a Multicultural, Postcolonial, and Feminist World,* ed. Uma Narayan and Sandra Harding (Bloomington: Indiana University Press, 2000), 177–88.

7. Bonnie Kettel, "Key Paths for Science and Technology," in *Missing Links: Gender Equity in Science and Technology for Development,* ed. Gender Working Group, United Nations Commission on Science and Technology for Development (Ottawa: International Development Research Centre, 1995).

8. For examples from the Global North, see Gloria Anzaldua, *Borderlands/La Frontera* (San Francisco: Spinsters/Aunt Lute, 1987); and Patricia Hill Collins, *Black Feminist Thought: Knowledge, Consciousness, and the Politics of Empowerment* (New York: Routledge, 1991).

9. An earlier version of the following section appeared in "Gender, Development, and Post-Enlightenment Philosophies of Science," *Hypatia* 13.3 (1998): 148–54.

10. Mies, *Patriarchy and Accumulation;* Shiva, *Staying Alive.*

11. Modernization ("development") in Europe took exactly this pattern with enclosure of the commons, which forced migration from rural to industrializing areas, creation of a proletariat, new marriage and inheritance laws favoring men, and so on. See, for example, Joan Kelly-Gadol, "The Social Relations of the Sexes: Methodological Implications of Women's History," *Signs: Journal of Women in Culture and Society* 1.4 (1976): 809–24.

12. Sparr, *Mortgaging Women's Lives.*

13. A few other accounts originating in the North have also drawn on the combined resources of feminist political economy and feminist environmental studies for their science studies projects. See Joni Seager, *Earth Follies: Coming to Feminist Terms with the Global Environmental Crisis* (New York: Routledge, 1993); Seager, "Rachel Carson Died of Breast Cancer: The Coming of Age of Feminist Environmentalism," *Signs: Journal of Women in Culture and Society* 28.3 (2003): 945–72; and Val Plumwood, *Feminism and the Mastery of Nature* (New York: Routledge, 1993). These overlap with and diverge from the GESD accounts.

14. Ruth Dixon-Mueller, "Women in Agriculture: Counting the Labor Force in Developing Countries," in *Beyond Methodology,* ed. J. Cook and M. M. Fonow (Bloomington: Indiana University Press, 1990), 226–47.

15. See chapter 7.

16. Braidotti et al., *Women and Sustainable Development;* Seager, *Earth Follies;* Seager, "Rachel Carson"; Shiva, *Staying Alive.*

17. Fredric Jameson, *Postmodernism; or, The Cultural Logic of Late Capitalism* (Durham: Duke University Press, 1991).

Chapter 7. The Political Unconscious of Western Science

1. Jameson examines the political unconscious of literature in *Political Unconscious: Narrative as a Socially Symbolic Act* (Ithaca: Cornell University Press, 1981).

2. See, for example, Seyla Benhabib, ed., *Democracy and Difference: Contesting the Boundaries of the Political* (Princeton: Princeton University Press, 1996); Stephen Gill, "Globalisation, Market Civilisation, and Disciplinary Neoliberalism," *Millennium: Journal of International Studies* 24.3 (1995): 399–423; and William I. Robinson, *Promoting Polyarchy: Globalization, U.S. Intervention, and Hegemony* (New York: Cambridge University Press, 1996).

3. See Rosi Braidotti, E. Charkiewicz, Sabina Hausler, and Saskia Wieringa, *Women, the Environment, and Sustainable Development* (Atlantic Highlands, N.J.: Zed Press, 1994); Wendy Harcourt, ed., *Feminist Perspectives on Sustainable Development* (London: Zed Press, 1994); Wolfgang Sachs, ed., *The Development Dictionary: A Guide to Knowledge as Power* (Atlantic Highlands, N.J.: Zed Press, 1992); Vandana Shiva, *Staying Alive: Women, Ecology, and Development* (London: Zed Press, 1989); and Pamela Sparr, ed., *Mortgaging Women's Lives: Feminist Critiques of Structural Adjustment* (London: Zed Press, 1994).

4. See David Noble, *A World without Women: The Christian Clerical Culture of Western Science* (New York: Alfred A. Knopf, 1992); and Alfred Sohn-Rethel, *Intellectual and Manual Labor* (London: Macmillan, 1978).

5. See, for example, Paul R. Gross and Norman Levitt, *Higher Superstition: The Academic Left and Its Quarrels with Science* (Baltimore: Johns Hopkins University Press, 1994); Paul R. Gross, Norman Levitt, and Martin W. Lewis, *The Flight from Science and Reason* (New York: New York Academy of Sciences, 1997); Sylvia Walby, "Against Epistemological Chasms: The Science Question in Feminism Revisited," *Signs: Journal of Women in Culture and Society* 26.2 (2001): 485–510; and Andrew Ross, ed., *The Science Wars* (Durham: Duke University Press, 1996).

6. See especially the discussion in chapter 5 of Karen Barad's and Joseph Rouse's analyses of how the nature that scientists encounter is already encultured by past and ongoing scientific practice. See chapter 9 for a discussion of relativism.

7. For a fuller discussion of this issue, see chap. 10 of my *Is Science Multicultural? Postcolonialisms, Feminisms, and Epistemologies* (Bloomington: Indiana University Press, 1998).

8. Here is sociologist Patricia Hill Collins on why African American women researchers who are studying African American women are unlikely to find useful positivist approaches to research:

Several requirements typify positivist methodological approaches. First, re-search methods generally require a distancing of the researcher from her or his "object" of study by defining the researcher as a "subject" with full human sub-jectivity and by objectifying the "object" of study. . . . A second requirement is the absence of emotion from the research process. . . . Third, ethics and val-ues are deemed inappropriate in the research process, either as the reason for scientific inquiry or as part of the research process itself. . . . Finally, adversar-ial debates, whether written or oral, become the preferred method of ascer-taining truth: the arguments that can withstand the greatest assault and sur-vive intact become the strongest truths.

Such criteria ask African-American women to objectify ourselves, devalue our emotional life, displace our motivations for furthering knowledge about Black women, and confront in an adversarial relationship those with more so-cial, economic, and professional power. It therefore seems unlikely that Black women would use a positivist epistemological stance in rearticulating a Black women's standpoint. (*Black Feminist Thought: Knowledge, Consciousness, and the Politics of Empowerment* [New York: Routledge, 1991], 205–6)

9. See Ulrich Beck, *Risk Society: Towards a New Modernity* (London: Sage, 1992), on this topic.

10. Sandra Harding, "Dysfunctional Universality Claims? Scientific, Epistemo-logical, and Political Issues," in *Is Science Multicultural?*

11. See John Dupre, *The Disorder of Things: Metaphysical Foundations for the Dis-unity of Science* (Cambridge: Harvard University Press, 1993); and Peter Galison and David J. Stump, *The Disunity of Science: Boundaries, Contexts, and Power* (Stanford: Stanford University Press, 1996).

12. Edward O. Wilson, *Consilience: The Unity of Knowledge* (New York: Alfred A. Knopf, 1998), 4, 266. Wilson is one of the original developers of sociobiology. See his *Sociobiology: The New Synthesis* (Cambridge: Harvard University Press, 1975).

13. For example, see Galison and Stump, *Disunity of Science.* I thank an anony-mous review of an earlier draft of this chapter for reminding me of the influential effects of Wilson's work and pointing out the complexity of current science stud-ies' thinking on this topic, to which I refer below.

14. For example, see ibid.; Nancy Cartwright, *Otto Neurath: Philosophy between Science and Politics* (New York: Cambridge University Press, 1996); and *Perspectives on Science* 7.3 (1999): 293–348, a symposium on the unity of science thesis.

15. See Ian Hacking, "The Disunities of the Sciences," in *Disunity of Science,* ed. Galison and Stump.

16. For a fuller development of these costs, see Harding, "Dysfunctional Univer-sality Claims?" in *Is Science Multicultural?* See also David J. Stump's catalog of costs of the unity thesis in "Afterword: New Directions in the Philosophy of Science Stud-ies," in *Disunity of Science,* ed. Galison and Stump.

17. See Vandana Shiva's discussion in *Monocultures of the Mind: Perspectives on*

Biodiversity and Biotechnology (New York and Penang, Malaysia: Zed Press and Third World Network, 1993).

18. See Joni Seager, *Earth Follies: Coming to Feminist Terms with the Global Environmental Crisis* (New York: Routledge, 1993).

19. See, for example, Gross and Levitt, *Higher Superstition;* and Gross, Levitt, and Lewis, *Flight from Science and Reason.*

20. See Ruth Hubbard, "Have Only Men Evolved?" in *Discovering Reality: Feminist Perspectives on Epistemology, Metaphysics, Methodology, and Philosophy of Science,* ed. Sandra Harding and Merrill Hintikka, 2d ed. (Dordrecht: Kluwer, 2003); and Anne Fausto-Sterling, *Myths of Gender: Biological Theories about Women and Men* (1985; 2d ed. New York: Basic Books, 1992).

21. Bruno Latour, *We Have Never Been Modern,* trans. Catherine Porter (Cambridge: Harvard University Press, 1993), 1–2.

22. See Michael Adas, *Machines as the Measure of Man* (Ithaca: Cornell University Press, 1989).

23. See Rachel Carson, *The Silent Spring* (Harmondsworth: Penguin, 1962).

24. See Nancy Hartsock on the "barracks mentality" ("The Feminist Standpoint: Developing the Ground for a Specifically Feminist Historical Materialism," in *Discovering Reality,* ed. Harding and Hintikka).

25. Michael Gibbons, Camille Limoges, Helga Nowotny, Simon Schwartzman, Peter Scott, and Martin Trow, *The New Production of Knowledge: The Dynamics of Science and Research in Contemporary Societies* (Thousand Oaks, Calif.: Sage, 1994).

26. See, for example, Paul Feyerabend, *Against Method* (London: New Left Books, 1975).

27. I notice, though, that graduate students have a more cynical perspective than do their professors on the extent to which this ideal actually functions in university labs today.

28. See Part 1.

29. See Sandra Harding, "Recovering Epistemological Resources: Strong Objectivity," chap. 8 of *Is Science Multicultural?*

30. See Sachs, *Development Dictionary.*

31. Jurgen Habermas, *Knowledge and Human Interests* (Boston: Beacon Press, 1971); Genevieve Lloyd, *The Man of Reason: "Male" and "Female" in Western Philosophy* (Minneapolis: University of Minnesota Press, 1984); Alison Jaggar, "Love and Knowledge: Emotion in Feminist Epistemology," in *Gender/Body/Knowledge,* ed. Susan Bordo and Alison Jaggar (New Brunswick, N.J.: Rutgers University Press, 1989); James Maffie, "To Walk in Balance: An Encounter between Contemporary Western Science and Conquest-Era Nahua Philosophy," in *Science and Other Cultures: Issues in Philosophies of Science and Technology,* ed. Robert Figueroa and Sandra Harding (New York: Routledge, 2003).

32. See Ashis Nandy, "Science as a Reason of State," in *Science, Hegemony, and Violence: A Requiem for Modernity,* ed. Nandy (New Delhi, India: Oxford University Press, 1990).

33. Latour, *We Have Never Been Modern;* Beck, *Risk Society.*

34. See earlier citations and the arguments of chapters 2–6.

Chapter 8. Are Truth Claims in Science Dysfunctional?

1. See W. V. O. Quine, "Two Dogmas of Empiricism," in *From a Logical Point of View,* by Quine (Cambridge: Harvard University Press, 1953).

2. See Pierre Duhem, *The Aim and Structure of Physical Theory,* trans. Philip Wiener (1906; reprint, Princeton: Princeton University Press, 1954).

3. See Karl Popper, *Conjectures and Refutations: The Growth of Scientific Knowledge,* 4th ed. (London: Routledge and Kegan Paul, 1972). See also Sandra G. Harding, ed., *Can Theories Be Refuted? Essays on the Duhem-Quine Thesis* (Dordrecht: Reidel, 1976).

4. Morris Kline makes similar arguments about mathematical principles in *Mathematics: The Loss of Certainty* (New York: Oxford University Press, 1980). See also Sal Restivo's development of a sociology of mathematical concepts, principles, and claims in *Mathematics in Society and History: Sociological Inquiries* (Dordrecht: Kluwer, 1992). David Bloor makes arguments for the historical contingency of mathematical claims in *Knowledge and Social Imagery* (London: Routledge and Kegan Paul, 1977).

5. See Paul Feyerabend, *Against Method* (London: New Left Books, 1975); and Thomas S. Kuhn, *The Structure of Scientific Revolutions,* 2d ed. (Chicago: University of Chicago Press, 1970).

6. See chapter 7 and Peter Galison and David J. Stump, *The Disunity of Science: Boundaries, Contexts, and Power* (Stanford: Stanford University Press, 1996).

7. We shall not here pursue important arguments against the first of these assumptions about the unity of nature that have emerged from within the natural sciences themselves as well as from philosophy, but see John Dupre, *The Disorder of Things: Metaphysical Foundations for the Disunity of Science* (Cambridge: Harvard University Press, 1993).

8. A. C. Crombie, *Styles of Scientific Thinking in the European Tradition* (London: Duckworth, 1994); Ian Hacking, "The Disunities of the Sciences," in *Disunity of Science,* ed. Galison and Stump.

9. Hacking, "Disunities of the Sciences," 68.

10. See Bloor, *Knowledge and Social Imagery;* Kline, *Mathematics;* and Restivo, *Mathematics in Society.*

11. The phrase is Andrew Pickering's in "Objectivity and the Mangle of Practice," in *Deconstructing and Reconstructing Objectivity,* a special issue of *Annals of Scholarship* 8 (1992): 409–25.

12. *Indigenous Knowledge and Development Monitor* (http://www.nuffic.ni/ciran /ikdm.html); Helaine Selin, ed., *Encyclopedia of the History of Science, Technology, and Medicine in Non-Western Cultures* (Dordrecht: Kluwer, 1997).

13. See, for example, Arturo Escobar, *Encountering Development: The Making and Unmaking of the Third World* (Princeton: Princeton University Press, 1995); Susantha Goonatilake, *Aborted Discovery: Science and Creativity in the Third World* (London:

Zed Press, 1984); David J. Hess, *Science and Technology in a Multicultural World: The Cultural Politics of Facts and Artifacts* (New York: Columbia University Press, 1995); Patrick Petitjean, Catherine Jami, and Anne Marie Moulin, eds., *Science and Empires: Historical Studies about Scientific Development and European Expansion* (Dordrecht: Kluwer, 1992); Wolfgang Sachs, ed., *The Development Dictionary: A Guide to Knowledge as Power* (Atlantic Highlands, N.J.: Zed Press, 1992); and chapters 2–6 above.

14. Thanks to Val Plumwood for pointing out to me this fourth assumption in the unity of science thesis.

15. Such insights are the beginning of the development of standpoint epistemologies. But they are only the beginning, not the end, since, for standpoint theorists, these insights must be used critically to "study up" and map the conceptual practices of power that enable the exploitation and domination of such groups. See chapter 5 and Sandra Harding, ed., *The Feminist Standpoint Theory Reader: Intellectual and Political Controversies* (New York: Routledge, 2003).

Chapter 9. Does the Threat of Relativism Deserve a Panic?

1. This is my way of formulating the problem. See my "Recovering Epistemological Resources: Strong Objectivity," chap. 8 of *Is Science Multicultural? Postcolonialisms, Feminisms, and Epistemologies* (Bloomington: Indiana University Press, 1998). For other valuable insights into how culture and politics inevitably enter the technical, cognitive core of scientific practice, see Karen Barad, "Posthumanist Performativity: Toward an Understanding of How Matter Comes to Matter," *Signs: Journal of Women in Culture and Society* 28.3 (2003): 801–32, and the discussion of her work in chapter 5; Donna Haraway, "Situated Knowledges," in *Simians, Cyborgs, and Women: The Reinvention of Nature* (New York: Routledge, 1991); and Evelyn Fox Keller, *Reflections on Gender and Science* (New Haven: Yale University Press, 1984).

2. See, for example, Haraway, "Situated Knowledges"; and Bruno Latour, *We Have Never Been Modern*, trans. Catherine Porter (Cambridge: Harvard University Press, 1993).

3. See the essays in Mario Biagioli, ed., *The Science Studies Reader* (New York: Routledge, 1999).

4. See Harding, *Is Science Multicultural?*

5. For example, see Steve Woolgar, ed., *Knowledge and Reflexivity* (Beverly Hills: Sage, 1988); Malcolm Ashmore, *The Reflexivity Thesis* (Chicago: University of Chicago Press, 1989); and H. M. Collins and Trevor Pinch, *The Golem: What Everyone Should Know about Science* (New York: Cambridge University Press, 1993).

6. Clifford Geertz, "Anti Anti-Relativism," in *Relativism: Interpretation and Confrontation*, ed. Michael Krausz (Notre Dame, Ind.: University of Notre Dame Press, 1989), 31–32.

7. See Paul Feyerabend, "Notes on Relativism," in *Farewell to Reason*, by Feyerabend (New York: Verso, 1987), 19–89; Lorraine Code, *What Can She Know?* (Ithaca:

NOTES TO PAGES 147–51 · 181

Cornell University Press, 1991); and Fredric Jameson, *"History and Class Consciousness* as an 'Unfinished Project,'" *Rethinking Marxism* 1 (1988): 49–72. The relevant section is reprinted in Sandra Harding, ed., *The Feminist Standpoint Theory Reader: Intellectual and Political Controversies* (New York: Routledge, 2003).

8. See note 1, this chapter.

9. Haraway, "Situated Knowledges," 187. See also, for example, Latour, *We Have Never Been Modern;* Ulrich Beck, *Risk Society: Towards a New Modernity* (London: Sage, 1992); and Ashis Nandy, ed., *Science, Hegemony, and Violence: A Requiem for Modernity* (New Delhi, India: Oxford University Press, 1990).

10. We can easily think of recent cases of such collective shifts in belief, such as AIDS activists' decision to develop alternative standards for remedy trials different from those of the National Institutes of Health and environmentalists' development of standards for pesticide use that differ from conventional ones.

11. See, for example, Krausz, *Relativism: Interpretation and Confrontation;* and Jack W. Meiland and Michael Krausz, eds., *Relativism: Cognitive and Moral* (Notre Dame, Ind.: University of Notre Dame Press, 1982).

12. For one of the most compelling arguments against the complete separability of words and objects, see Helen Verran, *Science and an African Logic* (Chicago: University of Chicago Press, 2001). Joseph Rouse's work is another good place to see the philosophic development of the kind of analysis I use here and in the following paragraphs. See his *Knowledge and Power: Toward a Political Philosophy of Science* (Ithaca: Cornell University Press, 1987); *Engaging Science: How to Understand Its Practices Philosophically* (Ithaca: Cornell University Press, 1996); and *How Scientific Practices Matter: Reclaiming Philosophical Naturalism* (Chicago: University of Chicago Press, 2002).

13. See Thomas S. Kuhn, *The Structure of Scientific Revolutions,* 2d ed. (Chicago: University of Chicago Press, 1970); Michel Foucault, *Discipline and Punish,* trans. Alan Sheridan (New York: Random House, 1977); Ian Hacking, *Representing and Intervening* (Cambridge: Cambridge University Press, 1983); Rouse, *Knowledge and Power, Engaging Science,* and *How Scientific Practices Matter;* and Karan Barad, "Meeting the Universe Halfway: Realism and Social Constructivism without Contradiction," in *Feminism, Science, and the Philosophy of Science,* ed. Jack Nelson and Lynn Hankinson Nelson (Dordrecht: Kluwer, 1996), "Agential Realism: Feminist Interventions in Understanding Scientific Practices," in *The Science Studies Reader,* ed. Biagioli, "Posthumanist Performativity," and *Meeting the Universe Halfway* (forthcoming).

14. See chapters 2–6.

15. It is odd to refer to poststructuralism as an epistemology or science movement. Yet it has been preoccupied with denying central philosophic assumptions of Enlightenment science, and it has explicitly been used to illuminating effect by some feminist science theorists—that is, by science studies scholars who have been part of such social movements. See, for example, Donna Haraway, *Primate Visions: Gender, Race, and Nature in the World of Modern Science* (New York: Routledge, 1989) and *Simians, Cyborgs, and Women.*

16. See Manuel Castells, *The Information Age: Economy, Society, and Culture*, vols. 1–3 (Oxford: Blackwell, 1996–2000); and Michael Gibbons, Camille Limoges, Helga Nowotny, Simon Schwartzman, Peter Scott, and Martin Trow, *The New Production of Knowledge: The Dynamics of Science and Research in Contemporary Societies* (Thousand Oaks, Calif.: Sage, 1994).

17. W. V. O. Quine, "Two Dogmas of Empiricism," in *From a Logical Point of View*, by Quine (Cambridge: Harvard University Press, 1953); Quine, *Word and Object* (Cambridge: MIT Press, 1960).

18. See Kuhn, *Structure of Scientific Revolutions*; and Foucault, *Discipline and Punish* and *The Birth of the Clinic*, trans. A. M. Sheridan Smith (New York: Vintage, 1994).

19. Some important contributors to or reporters of the postpositivist philosophies of science grounded in this understanding are Barry Barnes, *Interests and the Growth of Knowledge* (Boston: Routledge and Kegan Paul, 1977); David Bloor, *Knowledge and Social Imagery* (London: Routledge and Kegan Paul, 1977); Peter Galison and David J. Stump, *The Disunity of Science: Boundaries, Contexts, and Power* (Stanford: Stanford University Press, 1996); David J. Hess, *Science and Technology in a Multicultural World: The Cultural Politics of Facts and Artifacts* (New York: Columbia University Press, 1995); Bruno Latour, *The Pasteurization of France* (Cambridge: Harvard University Press, 1988) and *Science in Action* (Cambridge: Harvard University Press, 1987); Bruno Latour and Steve Woolgar, *Laboratory Life: The Social Construction of Scientific Facts* (Beverly Hills: Sage, 1979); Sal Restivo, *Mathematics in Society and History: Sociological Inquiries* (Dordrecht: Kluwer, 1992); Rouse, *Engaging Science*; John A. Schuster and Richard R. Yeo, eds., *The Politics and Rhetoric of Scientific Method: Historical Studies* (Dordrecht: D. Reidel, 1986); Steven Shapin, *A Social History of Truth* (Chicago: University of Chicago Press, 1994); Steven Shapin and Simon Shaffer, *Leviathan and the Air Pump* (Princeton: Princeton University Press, 1985); and Sharon Traweek, *Beamtimes and Life Times* (Cambridge: MIT Press, 1988). (I count the sociologists, historians, and ethnographers as full-fledged contributors to the new postpositivist philosophies of science.)

For feminist and postcolonial contributors and reporters, some of whom fully function in postpositivist Northern circles, but most of whom would have various qualms about identifying themselves as "postpositivist," see chapters 2–6.

20. As discussed in chapter 7, this still powerful though also discredited thesis holds that there is one world, one "truth" about it, and one and only one science capable in principle of representing that one true story. As political scientists point out, it also assumes that there is one and only one ideal kind of knower capable of producing such an account. This thesis united a social movement from the late nineteenth through mid-twentieth centuries that sought to reveal the singularity (in its reductionist versions) or, in most cases, just the "harmony" of diverse disciplinary investigations with each other. Thomas S. Kuhn's influential 1962 book significantly was the last to be published in the University of Chicago Press's Unity of Sciences series. See the essays in Galison and Stump, *Disunity of Science*, for histories and analyses of this movement.

21. See Harding, chaps. 4 and 6 of *Is Science Multicultural?* and chapters 2–6 above.

22. See the discussion of this point in N. Katherine Hayles, "Constrained Constructivism: Locating Scientific Inquiry in the Theater of Representation," in *Realism and Representation*, ed. George Levine (Madison: University of Wisconsin Press, 1993).

23. *Fallibilism* is frequently the term used to refer to this feature of empirical knowledge claims. Yet it is all too often invoked to admit that there is, of course, a rational possibility that the speaker's claims could at some point in the future reasonably be regarded as inadequate, but in the meantime such claims deserve to be regarded as robustly adequate or even true and to be firmly supported by solid empirical evidence and logical reasoning. See chapter 8.

24. See Verran, *African Logic*, for a fascinating account of how this is the case in both the West and Africa even for mathematical learning.

25. Of course, no culture is able to detect all of its cultural assumptions, as historians usefully point out. See Joseph Needham's illuminating account of the distinctively Western cultural values, including Christian beliefs, that shape purportedly value-neutral conceptual frameworks of modern Western sciences, in "Human Law and the Laws of Nature," in *The Grand Titration: Science and Society in East and West*, by Needham (Toronto: University of Toronto Press, 1969), 299–331.

26. See Haraway, "Situated Knowledges"; and Robert Proctor, *Value-Free Science? Purity and Power in Modern Knowledge* (Cambridge: Harvard University Press, 1991).

Bibliography

Adas, Michael. *Machines as the Measure of Man.* Ithaca: Cornell University Press, 1989.

Agarwal, Bina. "The Gender and Environment Debate: Lessons from India." *Feminist Studies* 18.1 (1993): 119–58.

American Association of University Women Educational Foundation. *Gender Gap: Where Schools Still Fail Our Children.* New York: Marlowe, 1998.

———. *Under the Microscope: A Decade of Gender Equity Projects in the Sciences.* Washington, D.C.: AAUW Educational Foundation, 2004.

Amin, Samir. *Eurocentrism.* New York: Monthly Review, 1989.

Anzaldua, Gloria. *Borderlands/La Frontera.* San Francisco: Spinsters/Aunt Lute, 1987.

Apffel-Marglin, Frederique, and Stephen A. Marglin. *Dominating Knowledge: Development, Culture, and Resistance.* Oxford: Clarendon Press, 1990.

Arditti, Rita, Renate Duelli-Klein, and Shelly Minden, eds. *Test-Tube Women: What Future for Motherhood?* Boston: Pandora Press, 1984.

Ashmore, Malcolm. *The Reflexivity Thesis.* Chicago: University of Chicago Press, 1989.

Bandyopadhyay, J., and Vandana Shiva. "Science and Control: Natural Resources and Their Exploitation." In *The Revenge of Athena: Science, Exploitation, and the Third World*, ed. Ziauddin Sardar. London: Mansell, 1988.

Barad, Karen. "Agential Realism: Feminist Interventions in Understanding Scientific Practices." In *The Science Studies Reader*, ed. Mario Biagioli. New York: Routledge, 1999.

———. "Getting Real: Technoscientific Practices and the Materialization of Reality." *Differences* 10.2 (1998): 87–128.

———. *Meeting the Universe Halfway.* Forthcoming.

———. "Meeting the Universe Halfway: Realism and Social Constructivism without Contradiction." In *Feminism, Science, and the Philosophy of Science*, ed. Jack Nelson and Lynn Hankinson Nelson. Dordrecht: Kluwer, 1996.

————. "Posthumanist Performativity: Toward an Understanding of How Matter Comes to Matter." *Signs: Journal of Women in Culture and Society* 28.3 (2003): 801–32.

Barker, Drucilla K. "Dualisms, Discourse, and Development." In *Decentering the Center: Philosophy for a Multicultural, Postcolonial, and Feminist World,* ed. Uma Narayan and Sandra Harding. Bloomington: Indiana University Press, 2000.

Barnes, Barry. *Interests and the Growth of Knowledge.* Boston: Routledge and Kegan Paul, 1977.

Basu, Amrita, Inderpal Grewal, Caren Kaplan, and Lisa Malkki, eds. *Globalization and Gender.* Special issue of *Signs: Journal of Women in Culture and Society* 26.4 (2001).

Beck, Ulrich. *Risk Society: Towards a New Modernity.* London: Sage, 1992.

————. *World Risk Society.* Oxford: Blackwell, 1999.

Benhabib, Seyla, ed. *Democracy and Difference: Contesting the Boundaries of the Political.* Princeton: Princeton University Press, 1996.

Berman, Morris. *The Reenchantment of the World.* Ithaca: Cornell University Press, 1981.

Bernal, Martin. *Black Athena: The Afroasiatic Roots of Classical Civilization.* Vol. 1. New Brunswick, N.J.: Rutgers University Press, 1987.

Biagioli, Mario, ed. *The Science Studies Reader.* New York: Routledge, 1999.

Blaut, J. M. *The Colonizer's Model of the World: Geographical Diffusionism and Eurocentric History.* New York: Guilford Press, 1993.

Bloor, David. *Knowledge and Social Imagery.* London: Routledge and Kegan Paul, 1977.

Bordo, Susan. *The Flight to Objectivity.* Albany: SUNY Press, 1987.

Boston Women's Health Collective. *Our Bodies, Ourselves: The Boston Women's Health Guide.* Boston: New England Free Press, 1970.

Braidotti, Rosi, E. Charkiewicz, Sabina Hausler, and Saskia Wieringa. *Women, the Environment, and Sustainable Development.* Atlantic Highlands, N.J.: Zed Press, 1994.

Brickhouse, Nancy. "Bringing in the Outsiders: Reshaping the Sciences of the Future." *Journal of Curriculum Studies* 26.4 (1994): 401–16.

————. "Embodying Science: A Feminist Perspective on Learning." *Journal of Research in Science Teaching* 38.3 (2001): 282–95.

————. "Feminism(s) and Science Education." *International Handbook of Science Education* (1998): 1067–81.

Brockway, Lucille H. *Science and Colonial Expansion: The Role of the British Royal Botanical Gardens.* New York: Academic Press, 1979.

Bug, Amy. "Has Feminism Changed Physics?" *Signs: Journal of Women in Culture and Society* 28.3 (2003): 881–900.

Bullard, Robert. *Dumping in Dixie: Race, Class, and Environmental Quality.* Boulder: Westview Press, 2000.

Carson, Rachel. *The Silent Spring.* Harmondsworth: Penguin, 1962.

Cartwright, Nancy. *Otto Neurath: Philosophy between Science and Politics.* New York: Cambridge University Press, 1996.

Castells, Manuel. *The Information Age: Economy, Society, and Culture.* Vols. 1–3. Oxford: Blackwell, 1996–2000.

Clarke, Adele E., and Virginia L. Olesen, eds. *Revisioning Women, Health, and Healing: Feminist, Cultural, and Technoscience Perspectives.* New York: Routledge, 1999.

Clough, Sharyn. *Beyond Epistemology: A Pragmatist Approach to Feminist Science Studies.* Lanham, Md.: Rowman and Littlefield, 2003.

Cockburn, Cynthia. *Machinery of Dominance: Women, Men, and Technical Know-How.* London: Pluto Press, 1985.

Code, Lorraine. *What Can She Know?* Ithaca: Cornell University Press, 1991.

Collins, H. M., and Trevor Pinch. *The Golem: What Everyone Should Know about Science.* New York: Cambridge University Press, 1993.

Collins, Patricia Hill. *Black Feminist Thought: Knowledge, Consciousness, and the Politics of Empowerment.* New York: Routledge, 1991.

Conkey, Margaret W. "Has Feminism Changed Archaeology?" *Signs: Journal of Women in Culture and Society* 28.3 (2003): 867–80.

Connell, Robert W. "Change among the Gatekeepers: Men, Masculinities, and Gender Equality in the Global Arena." *Signs: Journal of Women in Culture and Society* 30.3 (2005): 1801–26.

Cook, J., and M. M. Fonow, eds. *Beyond Methodology.* Bloomington: Indiana University Press, 1990.

Crombie, A. C. *Styles of Scientific Thinking in the European Tradition.* London: Duckworth, 1994.

Crosby, Alfred. *The Columbian Exchange: Biological and Cultural Consequences of 1492.* Westport, Conn.: Greenwood Press, 1972.

———. *Ecological Imperialism: The Biological Expansion of Europe.* Cambridge: Cambridge University Press, 1987.

Duhem, Pierre. *The Aim and Structure of Physical Theory.* Trans. Philip Wiener. 1906. Reprint, Princeton: Princeton University Press, 1954.

Dupre, John. *The Disorder of Things: Metaphysical Foundations for the Disunity of Science.* Cambridge: Harvard University Press, 1993.

———. "Metaphysical Disorder and Scientific Disunity." In *The Disunity of Science: Boundaries, Contexts, and Power,* ed. Peter Galison and David J. Stump. Stanford: Stanford University Press, 1996.

Escobar, Arturo. *Encountering Development: The Making and Unmaking of the Third World.* Princeton: Princeton University Press, 1995.

Etzkowitz, Henry, Carol Kemelgor, and Brian Uzzi. *Athena Unbound: The Advancement of Women in Science and Technology.* Cambridge: Cambridge University Press, 2000.

European Commission. *Waste of Talents: Turning Private Struggles into a Public Issue: Women and Science in the ENWISE Countries.* Luxembourg: European Communities, 2003.

Fausto-Sterling, Anne. "The Bare Bones of Sex: Part I—Sex and Gender." *Signs: Journal of Women in Culture and Society* 30.2 (2005): 1491–1528.

———. "The Five Sexes: Why Male and Female Are Not Enough." *Sciences* (March–April 1993): 20–24.

———. "Gender, Race, and Nation: The Comparative Anatomy of Hottentot Women in Europe I: 1815–1817." In *Deviant Bodies,* ed. Jennifer Terry and Jacqueline Urla. Bloomington: Indiana University Press, 1995.

———. "The Myth of Neutrality: Race, Sex, and Class in Science." *Radical Teacher* 19 (1981): 21–25.

———. *Myths of Gender: Biological Theories about Women and Men.* 1985. Reprint, New York: Basic Books, 1994.

———. "Race, Gender, and Science." *Transformations* 2.2 (1991): 4–12.

———. "Refashioning Race: DNA and the Politics of Health Care." *Differences: A Journal of Feminist Cultural Studies* 15.3 (2004): 1–37.

———. *Sexing the Body: Gender Politics and the Construction of Sexuality.* New York: Basic Books, 2000.

Feenberg, Andrew. "Technology in a Global World." In *Science and Other Cultures: Issues in Philosophies of Science and Technology,* ed. Robert Figueroa and Sandra Harding. New York: Routledge, 2003.

Feyerabend, Paul. *Against Method.* London: New Left Books, 1975.

———. "Notes on Relativism." In *Farewell to Reason,* by Paul Feyerabend, 19–89. New York: Verso, 1987.

Figueroa, Robert, and Sandra Harding, eds. *Science and Other Cultures: Issues in Philosophies of Science and Technology.* New York: Routledge, 2003.

Forman, Paul. "Behind Quantum Electronics: National Security as Bases for Physical Research in the U.S., 1940–1960." *Historical Studies in Physical and Biological Sciences* 18 (1987): 149–229.

Foucault, Michel. *The Birth of the Clinic.* Trans. A. M. Sheridan Smith. New York: Vintage, 1994.

———. *Discipline and Punish.* Trans. Alan Sheridan. New York: Random House, 1977.

———. *Power/Knowledge: Selected Interviews and Other Writings, 1972–77.* Trans. Colin Gordon, Leo Marshall, John Mepham, and Kate Soper. New York: Random House, 1980.

Fuller, Steve. Review of *The New Production of Knowledge: The Dynamics of Science and Research in Contemporary Societies,* by Michael Gibbons, Camille Limoges, Helga Nowotny, Simon Schwartzman, Peter Scott, and Martin Trow. *Sociology* 29.1 (1995): 159–67.

Galison, Peter. *How Experiments End.* Chicago: University of Chicago Press, 1987.

Galison, Peter, and David J. Stump, eds. *The Disunity of Science: Boundaries, Contexts, and Power.* Stanford: Stanford University Press, 1996.

Geertz, Clifford. "Anti Anti-Relativism." In *Relativism: Interpretation and Confrontation,* ed. Michael Krausz. Notre Dame, Ind.: University of Notre Dame Press, 1989.

Gender Working Group, U.N. Commission on Science and Technology for Development, ed. *Missing Links: Gender Equity in Science and Technology for Development.* Ottawa: International Development Research Centre, 1995.

Gibbons, Michael, Camille Limoges, Helga Nowotny, Simon Schwartzman, Peter Scott, and Martin Trow. *The New Production of Knowledge: The Dynamics of Science and Research in Contemporary Societies.* Thousand Oaks, Calif.: Sage, 1994.

Giddens, Anthony. *The Consequences of Modernity.* Stanford: Stanford University Press, 1990.

Gill, Stephen. "Globalisation, Market Civilisation, and Disciplinary Neoliberalism." *Millennium: Journal of International Studies* 24.3 (1995): 399–423.

Gilman, Sander L. *Difference and Pathology: Stereotypes of Sexuality, Race, and Madness.* Ithaca: Cornell University Press, 1985.

Godin, Benoit. "Writing Performative History: The New *New Atlantis?*" *Social Studies of Science* 28.3 (1998): 465–83.

Goonatilake, Susantha. *Aborted Discovery: Science and Creativity in the Third World.* London: Zed Press, 1984.

———. "A Project for Our Times." In *The Revenge of Athena: Science, Exploitation, and the Third World,* ed. Ziauddin Sardar. London: Mansell, 1988.

———. *Toward a Global Science: Mining Civilizational Knowledge.* Bloomington: Indiana University Press, 1998.

———. "The Voyages of Discovery and the Loss and Rediscovery of the 'Other's' Knowledge." *Impact of Science on Society,* no. 167 (1992): 241–64.

Gould, Stephen Jay. *The Mismeasure of Man.* New York: W. W. Norton, 1981.

Gowaty, Patricia Adair. "Sexual Natures: How Feminism Changed Evolutionary Biology." *Signs: Journal of Women in Culture and Society* 28.3 (2003): 901–22.

Gross, Paul R., and Norman Levitt. *Higher Superstition: The Academic Left and Its Quarrels with Science.* Baltimore: Johns Hopkins University Press, 1994.

Gross, Paul R., Norman Levitt, and Martin W. Lewis, eds. *The Flight from Science and Reason.* New York: New York Academy of Sciences, 1997.

Habermas, Jurgen. *Knowledge and Human Interests.* Boston: Beacon Press, 1971.

Hacking, Ian. "The Disunities of the Sciences." In *The Disunity of Science: Boundaries, Contexts, and Power,* ed. Peter Galison and David J. Stump. Stanford: Stanford University Press, 1996.

———. *Representing and Intervening.* Cambridge: Cambridge University Press, 1983.

Hammonds, Evelynn. *The Logic of Difference: A History of Race in Science and Medicine in the United States.* Forthcoming.

Hammonds, Evelynn, and Banu Subramaniam. "A Conversation on Feminist Science Studies." *Signs: Journal of Women in Culture and Society* 28.3 (2003): 923–44.

Haraway, Donna. *The Companion Species Manifesto: Dogs, People, and Significant Otherness.* Chicago: Prickly Paradigm Press, 2003.

———. *Modest_Witness@Second_Millennium.FemaleMan_Meets_OncoMouse: Feminism and Technoscience.* New York: Routledge, 1997.

———. *Primate Visions: Gender, Race, and Nature in the World of Modern Science.* New York: Routledge, 1989.

———. *Simians, Cyborgs, and Women: The Reinvention of Nature.* New York: Routledge, 1991.

Harcourt, Wendy, ed. *Feminist Perspectives on Sustainable Development.* London: Zed Press, 1994.

Harding, Sandra. "After the Neutrality Ideal: Science, Politics, and 'Strong Objectivity.'" *Social Research* 59 (1992): 567–87.

———. "Is Modern Science an Ethnoscience?" In *Sociology of the Sciences Yearbook,* ed. T. Shinn, J. Spaapen, and Raoul Waast. Dordrecht: Kluwer, 1996.

———. "Is Science Multicultural? Challenges, Resources, Opportunities, Uncertainties." *Configurations* 2.2 (1994): 301–30. First published in David Theo Goldberg, ed., *Multiculturalism: A Reader* (London: Blackwell's, 1994).

———. *Is Science Multicultural? Postcolonialisms, Feminisms, and Epistemologies.* Bloomington: Indiana University Press, 1998.

———. "Multicultural and Global Feminist Philosophies of Science: Resources and Challenges." In *Feminism, Science, and the Philosophy of Science,* ed. Jack Nelson and Lynn Hankinson Nelson. Dordrecht: Kluwer, 1996.

———. "Rethinking Standpoint Epistemology: What Is 'Strong Objectivity'?" In *Feminist Epistemologies,* ed. Linda Alcoff and Elizabeth Potter. New York: Routledge, 1992.

———. *The Science Question in Feminism.* Ithaca: Cornell University Press, 1986.

———. *Whose Science? Whose Knowledge?* Ithaca: Cornell University Press, 1991.

———. "Why Has the Sex/Gender System Become Visible Only Now?" In *Discovering Reality: Feminist Perspectives on Epistemology, Metaphysics, Methodology, and Philosophy of Science,* ed. Sandra Harding and Merrill Hintikka. Dordrecht: Reidel, 1983.

———. "Women and Science in Historical Context." *NWSA Journal* 5.1 (1993): 49–55.

———, ed. *Can Theories Be Refuted? Essays on the Duhem-Quine Thesis.* Dordrecht: Reidel, 1976.

———. *Feminism and Methodology: Social Science Issues.* Bloomington: Indiana University Press, 1987.

———. *The Feminist Standpoint Theory Reader: Intellectual and Political Controversies.* New York: Routledge, 2004.

———. *The "Racial" Economy of Science: Toward a Democratic Future.* Bloomington: Indiana University Press, 1993.

Harding, Sandra, and Merrill Hintikka, eds. *Discovering Reality: Feminist Perspectives on Epistemology, Metaphysics, Methodology, and Philosophy of Science.* 2d ed. Dordrecht: Kluwer, 2003.

Harding, Sandra, and Elizabeth McGregor. "The Gender Dimension of Science and Technology." In *UNESCO World Science Report,* ed. Howard J. Moore. Paris: UNESCO, 1996.

Harding, Sandra, and Kate Norberg, eds. *New Feminist Approaches to Social Science Methodology.* Special issue of *Signs: Journal of Women in Culture and Society* 30.4 (2005).

Harding, Sandra, and Jean O'Barr, eds. *Sex and Scientific Inquiry.* Chicago: University of Chicago Press, 1987.

Hardt, Michael, and Kathi Weeks. *The Jameson Reader.* Oxford: Blackwell, 2000.

Hartsock, Nancy. "The Feminist Standpoint: Developing the Ground for a Specifi-cally Feminist Historical Materialism." In *Discovering Reality: Feminist Perspectives on Epistemology, Metaphysics, Methodology, and Philosophy of Science,* ed. Sandra Harding and Merrill Hintikka. Dordrecht: Reidel, (1983) 2003.

Harvey, Elizabeth D. "Anatomies of Rapture: Clitoral Politics/Medical Blazons." *Signs: Journal of Women in Culture and Society* 27.2 (2002): 315–46.

Hayles, N. Katherine. "Constrained Constructivism: Locating Scientific Inquiry in the Theater of Representation." In *Realism and Representation,* ed. George Levine. Madison: University of Wisconsin Press, 1993.

Headrick, Daniel R., ed. *The Tools of Empire: Technology and European Imperialism in the Nineteenth Century.* New York: Oxford University Press, 1981.

Hess, David J. *Science and Technology in a Multicultural World: The Cultural Politics of Facts and Artifacts.* New York: Columbia University Press, 1995.

Hine, Darlene Clark. "Co-laborers in the Work of the Lord: Nineteenth-Century Black Women Physicians." In *"Send Us a Lady Physician": Women Doctors in America, 1835–1920,* ed. Ruth J. Abram. New York: W. W. Norton, 1985.

Horton, Robin. "African Traditional Thought and Western Science." Parts 1 and 2. *Africa* 37 (1967): 50–71, 155–87.

Hubbard, Ruth. "Science, Power, Gender: How DNA Became the Book of Life." *Signs: Journal of Women in Culture and Society* 28.3 (2003): 791–800.

Hutchins, Edwin. *Cognition in the Wild.* Cambridge: MIT Press, 1995.

Indigenous Knowledge and Development Monitor. http://www.nuffic.ni/ciran/ikdm .html.

Jacob, Margaret. *The Cultural Meanings of the Scientific Revolution.* New York: Alfred A. Knopf, 1988.

Jaggar, Alison. "Feminist Politics and Epistemology: Justifying Feminist Theory." Chap. 11 of *Feminist Politics and Human Nature.* Totowa, N.J.: Rowman and Allenheld, 1983.

———. "Love and Knowledge: Emotion in Feminist Epistemology." In *Gender/ Body/Knowledge,* ed. Susan Bordo and Alison Jaggar. New Brunswick, N.J.: Rutgers University Press, 1989.

Jameson, Fredric. *"History and Class Consciousness* as an 'Unfinished Project.'" *Rethinking Marxism* 1 (1988): 49–72.

———. *The Political Unconscious: Narrative as a Socially Symbolic Act.* Ithaca: Cornell University Press, 1981.

———. *Postmodernism; or, The Cultural Logic of Late Capitalism.* Durham: Duke University Press, 1991.

Jensen, A. R. *Bias in Mental Testing.* New York: Free Press, 1980.

Jones, James H. *Bad Blood: The Tuskegee Syphilis Experiment.* New York: Free Press, 1981.

Jordan, Winthrop. *White Over Black: American Attitudes toward the Negro, 1550–1812.* Chapel Hill: University of North Carolina Press, 1968.

Joseph, George Gheverghese. *The Crest of the Peacock: Non-European Roots of Mathematics.* New York: I. B. Tauris, 1991.

Keller, Evelyn Fox. *A Feeling for the Organism.* San Francisco: Freeman, 1983.

———. *Reflections on Gender and Science.* New Haven: Yale University Press, 1984.

———. *Secrets of Life, Secrets of Death: Essays on Language, Gender, and Science.* New York: Routledge, 1992.

Kelly, Alison, ed. *The Missing Half: Girls and Science Education.* Manchester: Manchester University Press, 1981.

———. *Science for Girls?* Philadelphia: Open University Press, 1987.

Kelly-Gadol, Joan. "The Social Relations of the Sexes: Methodological Implications of Women's History." *Signs: Journal of Women in Culture and Society* 1.4 (1976): 809–24.

Kettel, Bonnie. "Key Paths for Science and Technology." In *Missing Links: Gender Equity in Science and Technology for Development,* ed. Gender Working Group, United Nations Commission on Science and Technology for Development. Ottawa: International Development Research Centre, 1995.

Khor, Kok Peng. "Science and Development: Underdeveloping the Third World." In *The Revenge of Athena: Science, Exploitation, and the Third World,* ed. Ziauddin Sardar. London: Mansell, 1988.

Kleinman, Daniel Lee, and Steven P. Vallas. "Science, Capitalism, and the Rise of the 'Knowledge Worker': The Changing Structure of Knowledge Production in the United States." *Theory and Society* 30 (2001): 451–92.

Kline, Morris. *Mathematics: The Loss of Certainty.* New York: Oxford University Press, 1980.

Knorr-Cetina, Karin. *The Manufacture of Knowledge: An Essay on the Constructivist and Contextual Nature of Science.* New York: Pergamon, 1981.

Kochhar, R. K. "Science in British India." Parts 1 and 2. *Current Science* (India) 63.11 (1992): 689–94; 64.1 (1993): 55–62.

Krausz, Michael, ed. *Relativism: Interpretation and Confrontation.* Notre Dame, Ind.: University of Notre Dame Press, 1989.

Kuhn, Thomas S. *The Structure of Scientific Revolutions.* 2d ed. Chicago: University of Chicago Press, 1970.

Kumar, Deepak. "Problems in Science Administration: A Study of the Scientific Surveys in British India, 1757–1900." In *Science and Empires: Historical Studies about Scientific Development and European Expansion,* ed. Patrick Petitjean, Catherine Jami, and Anne Marie Moulin. Dordrecht: Kluwer, 1992.

———. *Science and Empire: Essays in Indian Context (1700–1947).* Delhi, India: Anamika Prakashan and National Institute of Science, Technology, and Development, 1991.

Lach, Donald F. *Asia in the Making of Europe.* Vol. 2. Chicago: University of Chicago Press, 1977.

Latour, Bruno. *The Pasteurization of France.* Cambridge: Harvard University Press, 1988.

————. *Science in Action*. Cambridge: Harvard University Press, 1987.

————. *We Have Never Been Modern*. Trans. Catherine Porter. Cambridge: Harvard University Press, 1993.

Latour, Bruno, and Steve Woolgar. *Laboratory Life: The Social Construction of Scientific Facts*. Beverly Hills: Sage, 1979.

Lewontin, Richard C., Steven Rose, and Leon J. Kamin. *Not in Our Genes*. New York: Pantheon, 1984.

Lionnet, Francoise, Obioma Nnaemeka, Susan Perry, and Celeste Schenk, eds. *Development Cultures: New Environments, New Realities, New Strategies*. Special issue of *Signs: Journal of Women in Culture and Society* 29.2 (2004).

Lloyd, Genevieve. *The Man of Reason: "Male" and "Female" in Western Philosophy*. Minneapolis: University of Minnesota Press, 1984.

Longino, Helen. *Science as Social Knowledge*. Princeton: Princeton University Press, 1990.

Maddox, Brenda. *Rosalind Franklin: The Dark Lady of DNA*. New York: Harper-Collins, 2002.

Maffie, James. "To Walk in Balance: An Encounter between Contemporary Western Science and Conquest-Era Nahua Philosophy." In *Science and Other Cultures: Issues in Philosophies of Science and Technology*, ed. Robert Figueroa and Sandra Harding. New York: Routledge, 2003.

Maines, Rachel P. *The Technology of Orgasm: "Hysteria," the Vibrator, and Women's Sexual Satisfaction*. Baltimore: Johns Hopkins University Press, 1999.

Malinowski, Bronislaw. *Magic, Science, and Religion, and Other Essays*. 1925. Reprint, Garden City, N.Y.: Doubleday Anchor, 1948.

Manning, Kenneth R. *Black Apollo of Science: The Life of Ernest Everett Just*. Oxford: Oxford University Press, 1983.

Marchessault, Janine, and Kim Sawchuk. *Wild Science: Reading Feminism, Medicine, and the Media*. London: Routledge, 2000.

Maxwell, Linda, Karen Slavin, and Kerry Young, eds. *Gender and Research*. Brussels: European Commission, 2002.

Mayberry, Maralee, Banu Subramaniam, and Lisa H. Weasel, eds. *Feminist Science Studies: A New Generation*. New York: Routledge, 2001.

McClellan, James E. *Colonialism and Science: Saint Domingue in the Old Regime*. Baltimore: Johns Hopkins University Press, 1992.

Meiland, Jack W., and Michael Krausz, eds. *Relativism: Cognitive and Moral*. Notre Dame, Ind.: University of Notre Dame Press, 1982.

Merchant, Carolyn. *The Death of Nature: Women, Ecology, and the Scientific Revolution*. New York: Harper and Row, 1980.

Mies, Maria. *Patriarchy and Accumulation on a World Scale: Women in the International Division of Labor*. Atlantic Highlands, N.J.: Zed Press, 1986.

Moore, Jason W. "The Crisis of Feudalism: An Environmental History." *Organization and Environment* 15.3 (2002): 301–22.

Murata, Junichi. "Creativity of Technology and the Modernization Process of Japan." In *Science and Other Cultures: Issues in Philosophies of Science and Technology*, ed. Robert Figueroa and Sandra Harding. New York: Routledge, 2003.

Murray, Margaret A. M. *Women Becoming Mathematicians: Creating a Professional Identity in Post–World War II America*. Cambridge: MIT Press, 2000.

Nader, Laura, ed. *Naked Science: Anthropological Inquiry into Boundaries, Power, and Knowledge*. New York: Routledge, 1996.

Nanda, Meera. *Prophets Facing Backward: Postmodern Critiques of Science and Hindu Nationalism in India*. New Brunswick, N.J.: Rutgers University Press, 2004.

Nandy, Ashis, ed. *Science, Hegemony, and Violence: A Requiem for Modernity*. New Delhi, India: Oxford University Press, 1990.

Narayan, Uma. "The Project of a Feminist Epistemology; Perspectives from a Nonwestern Feminist." In *Gender/Body/Knowledge*, ed. Susan Bordo and Alison Jaggar. New Brunswick, N.J.: Rutgers University Press, 1989.

Nasr, Seyyed Hossein. "Islamic Science, Western Science: Common Heritage, Diverse Destinies." In *The Revenge of Athena: Science, Exploitation, and the Third World*, ed. Ziauddin Sardar. London: Mansell, 1988.

National Science Foundation. *New Formulas for America's Workforce: Girls in Science and Engineering*. Washington, D.C.: National Science Foundation Education and Human Resources Directorate, 2004.

Needham, Joseph. *The Grand Titration: Science and Society in East and West*. Toronto: University of Toronto Press, 1969.

———. *Science and Civilisation in China*. 7 vols. Cambridge: Cambridge University Press, 1954–.

Nelson, Jack, and Lynn Hankinson Nelson. *Feminism, Science, and the Philosophy of Science*. Dordrecht: Kluwer, 1996.

Noble, David. *The Religion of Technology*. New York: Alfred A. Knopf, 1995.

———. *A World without Women: The Christian Clerical Culture of Western Science*. New York: Alfred A. Knopf, 1992.

Oldenziel, Ruth. *Making Technology Masculine: Women, Men, and the Machine in America, 1880–1945*. Ann Arbor and Amsterdam: University of Michigan Press and Amsterdam University Press, 1999.

Oldenziel, Ruth, Annie Canel, and Karin Zachmann, eds. *Building Bridges, Crossing Boundaries: Comparing the History of Women Engineers*. London: Harwood, 2000.

Omi, Michael, and Howard Winant. *Racial Formation in the United States*. New York: Routledge, 1994.

Pearson, Willie, Jr. *Black Scientists, White Society, and Colorless Science: A Study of Universalism in American Science*. Milwood, N.Y.: Associated Faculty Press, 1985.

Petitjean, Patrick, Catherine Jami, and Anne Marie Moulin, eds. *Science and Empires: Historical Studies about Scientific Development and European Expansion*. Dordrecht: Kluwer, 1992.

Plumwood, Val. *Feminism and the Mastery of Nature*. New York: Routledge, 1993.

Popper, Karl. *Conjectures and Refutations: The Growth of Scientific Knowledge.* 4th ed. London: Routledge and Kegan Paul, 1972.

Potter, Elizabeth. *Gender and Boyle's Law of Gases.* Bloomington: Indiana University Press, 2001.

Proctor, Robert. *Racial Hygiene: Medicine under the Nazis.* Cambridge: Harvard University Press, 1988.

———. *Value-Free Science? Purity and Power in Modern Knowledge.* Cambridge: Harvard University Press, 1991.

Quine, W. V. O. "Two Dogmas of Empiricism." In *From a Logical Point of View,* by W. V. O. Quine. Cambridge: Harvard University Press, 1953.

———. *Word and Object.* Cambridge: MIT Press, 1960.

Raffensperger, Carolyn, and Joel Tickner, eds. *Protecting Public Health and the Environment: Implementing the Precautionary Principle.* Washington, D.C.: Island Press, 1999.

Ravetz, Jerome. *Scientific Knowledge and Its Social Problems.* New York: Oxford University Press, 1971.

Reingold, Nathan, and Marc Rothenberg, eds. *Scientific Colonialism: Cross-Cultural Comparisons.* Washington, D.C.: Smithsonian Institution Press, 1987.

Restivo, Sal. *Mathematics in Society and History: Sociological Inquiries.* Dordrecht: Kluwer, 1992.

Robinson, William I. *Promoting Polyarchy: Globalization, U.S. Intervention, and Hegemony.* New York: Cambridge University Press, 1996.

Rose, Hilary. "Hand, Brain, and Heart: A Feminist Epistemology for the Natural Sciences." *Signs: Journal of Women in Culture and Society* 9.1 (1983): 73–90.

Rose, Hilary, and Steven Rose. "The Incorporation of Science." In *The Political Economy of Science: Ideology of/in the Natural Sciences,* ed. Hilary Rose and Steven Rose. London: Macmillan, 1976.

Ross, Andrew, ed. *The Science Wars.* Durham: Duke University Press, 1996.

Rosser, Sue V. "Female Friendly Science: Including Women in Curricular Content and Pedagogy in Science." *Journal of General Education* 42.3 (1993): 191–220.

———. *Teaching Science and Health from a Feminist Perspective: A Practical Guide.* New York: Pergamon, 1986.

———. *Women, Science, and Society: The Crucial Union.* New York: Teacher's College Press, 2000.

Rossiter, Margaret. *Women Scientists in America: Before Affirmative Action.* Baltimore: Johns Hopkins University Press, 1995.

———. *Women Scientists in America: Struggles and Strategies to 1940.* Baltimore: Johns Hopkins University Press, 1982.

Rouse, Joseph. "Barad's Feminist Naturalism." *Hypatia: A Journal of Feminist Philosophy* 19.1 (2004): 142–61.

———. *Engaging Science: How to Understand Its Practices Philosophically.* Ithaca: Cornell University Press, 1996.

————. *How Scientific Practices Matter: Reclaiming Philosophical Naturalism.* Chicago: University of Chicago Press, 2002.

————. *Knowledge and Power: Toward a Political Philosophy of Science.* Ithaca: Cornell University Press, 1987.

Sabra, I. A. "The Scientific Enterprise." In *The World of Islam,* ed. B. Lewis. London: Thames and Hudson, 1976.

Sachs, Wolfgang, ed. *The Development Dictionary: A Guide to Knowledge as Power.* Atlantic Highlands, N.J.: Zed Press, 1992.

Said, Edward. *Orientalism.* New York: Pantheon, 1978.

Sardar, Ziauddin, ed. *The Revenge of Athena: Science, Exploitation, and the Third World.* London: Mansell, 1988.

Sayre, Anne. *Rosalind Franklin and DNA.* New York: W. W. Norton, 1975.

Schiebinger, Londa. *Has Feminism Changed Science?* Cambridge: Harvard University Press, 1999.

————. *The Mind Has No Sex? Women in the Origins of Modern Science.* Cambridge: Harvard University Press, 1989.

————. *Nature's Body: Gender in the Making of Modern Science.* Boston: Beacon Press, 1993.

Schiebinger, Londa, Angela N. H. Creager, and Elizabeth Lunbeck, eds. *Feminism in Twentieth-Century Science, Technology, and Medicine.* Chicago: University of Chicago Press, 1991.

Schilpp, P. A., ed. *The Philosophy of Rudolf Carnap.* La Salle, Ill.: Open Court, 1963.

Schuster, John A., and Richard R. Yeo, eds. *The Politics and Rhetoric of Scientific Method: Historical Studies.* Dordrecht: D. Reidel, 1986.

Schwartz, Charles. "Political Structuring of the Institutions of Science." In *Naked Science: Anthropological Inquiry into Boundaries, Power, and Knowledge,* ed. Laura Nader, 148–59. New York: Routledge, 1996.

Seager, Joni. *Earth Follies: Coming to Feminist Terms with the Global Environmental Crisis.* New York: Routledge, 1993.

————. "Rachel Carson Died of Breast Cancer: The Coming of Age of Feminist Environmentalism." *Signs: Journal of Women in Culture and Society* 28.3 (2003): 945–72.

Selin, Helaine, ed. *Encyclopedia of the History of Science, Technology, and Medicine in Non-Western Cultures.* Dordrecht: Kluwer, 1997.

Sen, Gita, and Caren Grown. *Development Crises and Alternative Visions: Third World Women's Perspectives.* New York: Monthly Review Press, 1987.

Shapin, Steven. *A Social History of Truth.* Chicago: University of Chicago Press, 1994.

Shapin, Steven, and Simon Schaffer. *Leviathan and the Air Pump.* Princeton: Princeton University Press, 1985.

Shiva, Vandana. *Close to Home: Women Reconnect Ecology, Health, and Development Worldwide.* Philadelphia: New Society, 1994.

————. *Monocultures of the Mind: Perspectives on Biodiversity and Biotechnology.* New York and Penang, Malaysia: Zed Press and Third World Network, 1993.

————. *Staying Alive: Women, Ecology, and Development.* London: Zed Press, 1989.

———. *Stolen Harvest: The Hijacking of the Global Food Supply.* Cambridge, Mass.: South End Press, 2000.

Smith, Dorothy E. "Comment on Hekman's 'Truth and Method: Feminist Standpoint Theory Revisited.'" *Signs: Journal of Women in Culture and Society* 22.2 (1997): 392–98.

———. *The Conceptual Practices of Power: A Feminist Sociology of Knowledge.* Boston: Northeastern University Press, 1990.

———. *The Everyday World as Problematic: A Sociology for Women.* Boston: Northeastern University Press, 1987.

———. *Texts, Facts, and Femininity: Exploring the Relations of Ruling.* New York: Routledge, 1990.

———. *Writing the Social: Critique, Theory, and Investigations.* Toronto: University of Toronto Press, 1999.

Smith, Linda Tuhiwai. *Decolonizing Methodologies: Research and Indigenous Peoples.* London: University of Otago Press, 1999.

Snow, C. P. *The Two Cultures: And a Second Look.* 1959. Reprint, Cambridge: Cambridge University Press, 1964.

Sohn-Rethel, Alfred. *Intellectual and Manual Labor.* London: Macmillan, 1978.

Sokal, Alan, and Jean Bricmont. *Fashionable Nonsense: Postmodern Intellectuals' Abuse of Science.* New York: Picador USA, 1998.

Spanier, Bonnie. *Im/partial Science: Gender Ideology in Molecular Biology.* Bloomington: Indiana University Press, 1995.

Sparr, Pamela, ed. *Mortgaging Women's Lives: Feminist Critiques of Structural Adjustment.* London: Zed Press, 1994.

Stein, Barbara R. *On Her Own Terms: Annie Montague Alexander and the Rise of Science in the American West.* Berkeley and Los Angeles: University of California Press, 2001.

Steingraber, Sandra. *Living Downstream.* New York: Vintage, 1997.

Stepan, Nancy Leys. *The Idea of Race in Science: Great Britain, 1800–1960.* London: Macmillan, 1982.

———. "Race and Gender: The Role of Analogy in Science." *Isis* 77 (1986).

Subramaniam, Banu. *A Question of Variation: Race, Gender, and the Practice of Science.* Forthcoming.

Third World Network. *Modern Science in Crisis: A Third World Response.* Penang, Malaysia: Third World Network, 1988.

Tobach, Ethel, and Betty Rosoff, eds. *Genes and Gender.* Vols. 1–4. New York: Gordian Press, 1978–1984.

Todorov, Tzvetan. *The Conquest of America: The Question of the Other.* Trans. Richard Howard. New York: Harper and Row, 1984.

Traweek, Sharon. *Beamtimes and Life Times.* Cambridge: MIT Press, 1988.

Tuana, Nancy. "Coming to Understand: Orgasm and the Epistemology of Ignorance." *Hypatia: A Journal of Feminist Philosophy* 19.1 (2004): 194–232.

Van den Daele, Werner. "The Social Construction of Science." In *The Social Pro-*

duction of Scientific Knowledge, ed. E. Mendelsohn, P. Weingart, and R. Whitley. Dordrecht: Reidel, 1977.

Van Sertima, Ivan. *Blacks in Science: Ancient and Modern.* New Brunswick, N.J.: Transaction Books, 1986.

Verran, Helen. *Science and an African Logic.* Chicago: University of Chicago Press, 2001.

Wajcman, Judy. *Feminism Confronts Technology.* University Park: Pennsylvania State University Press, 1991.

Walby, Sylvia. "Against Epistemological Chasms: The Science Question in Feminism Revisited." *Signs: Journal of Women in Culture and Society* 26.2 (2001): 485–510.

Watson, James D. *The Double Helix.* New York: Atheneum, 1968.

Weatherford, Jack. *Indian Givers: What the Native Americans Gave to the World.* New York: Crown, 1988.

Wellman, David. *Portraits of White Racism.* New York: Cambridge University Press, 1977.

Williams, Kathleen Broome. *Improbable Warriors: Women Scientists and the U.S. Navy in World War II.* Annapolis, Md.: Naval Institute Press, 2001.

Wilson, Anna. "Sexing the Hyena: Intraspecies Readings of the Female Phallus." *Signs: Journal of Women in Culture and Society* 28.3 (2003): 755–90.

Wilson, Edward O. *Consilience: The Unity of Knowledge.* New York: Alfred A. Knopf, 1998.

———. *Sociobiology: The New Synthesis.* Cambridge: Harvard University Press, 1975.

Wiredu, J. E. "How Not to Compare African Thought with Western Thought." In *African Philosophy,* ed. Richard Wright. 3d ed. Lanham, Md.: University Press of America, 1984.

Wolf, Eric. *Europe and the Peoples without a History.* Berkeley and Los Angeles: University of California Press, 1984.

Woolgar, Steve. *Science: The Very Idea.* New York: Tavistock, 1988.

———, ed. *Knowledge and Reflexivity.* Beverly Hills: Sage, 1988.

Wyer, Mary, Mary Barbercheck, Donna Giesman, Hatice Örün Öztürk, and Marta Wayne. *Women, Science, and Technology: A Reader in Feminist Science Studies.* New York: Routledge, 2001.

Wylie, Alison. "The Constitution of Archaeological Evidence: Gender Politics and Science." In *The Disunity of Science: Boundaries, Contexts, and Power,* ed. Peter Galison and David J. Stump. Stanford: Stanford University Press, 1996.

———. *Gender, Politics, and Scientific Archaeology: The Challenge of "Gender Research" in Archaeology.* New York: Blackwell, 2004.

———. "Why Standpoint Matters." In *Science and Other Cultures: Issues in Philosophies of Science and Technology,* ed. Robert Figueroa and Sandra Harding. New York: Routledge, 2003.

Yates, Frances. *Giordano Bruno and the Hermetic Tradition.* New York: Vintage, 1969.

Zilsel, Edgar. "The Sociological Roots of Science." *American Journal of Sociology* 47 (1942).

Index

accountability, scientific, 10, 31, 46, 96, 107; democratic ideals in, 118–19; feminist research on, 81; of Northern science, 65
American Association for the Advancement of Science, 73
Amin, Samir, 165n9
astronomy: cultural meaning of, 37; non-Western, 36–37
Austin, J. L., 94

Barad, Karen, 90, 93–94, 96; on encultured nature, 176n6; on intervention, 150
Beck, Ulrich, 9
belief systems: about nature, 145; about social relations, 145; culturally embedded, 148, 154; empirical falsification of, 135; justification for, 148; networks of, 151, 153; Northern scientific, 139; objective, 149; rational, 150; shifts in, 181n10; transcultural, 145–46
biologists, feminist, 72, 124
biology: Eurocentrism in, 61–62; gender studies in, 73; racial difference in, 22, 23, 24; white supremacy in, 17
biomedicine: Chinese, 58; Western *versus* traditional, 28–29, 155. *See also* medicine
Bloor, David, 179n4
botany: in colonial India, 42; ethno-, 58
Boyle, Robert, 77
Brickhouse, Nancy, 78, 162n10, 170n30
Butler, Judith, 94

Carson, Rachel, 54, 127
Carver, George Washington, 26
chemistry: cultural understanding of, 33; Eurocentrism in, 61–62
China: biomedicine of, 58; knowledge traditions of, 40; science of, 34, 37–38, 45; use of Western medicine, 57
class: in approach to science, 52; and gender, 68; role in racial classification, 23; role of science in, 4; in scientific decision making, 120; standpoint theory and, 89
Clough, Sharyn, 94, 171n3, 174n25
Code, Lorraine, 147
Collins, Patricia Hill, 176n8
colonialism: end of, 58, 151; marriage policies under, 23; racism and, 29; skills learned under, 60. *See also* Postcolonialism
corporations, transnational, 56; and culturally neutral science, 129; in Global South, 115; profiteering by, 101
craniology, 19, 20
Crick, Francis, 71
Crombie, A. C., 138
cultural diversity: cognitive, 129; reduction in, 55
culture: binaries in, 12; effect of scientific rationality on, 130; gender-segregated, 167n6; in growth of knowledge, 153–54; influence on science, 30, 43–48, 81, 82, 89–90, 135, 143, 154, 164n28, 180n1; in philosophies of science, 82, 93; in production of knowledge,

SANDRA HARDING is a professor of philosophy in the Graduate School of Education and Information Studies at the University of California at Los Angeles. Harding, known for her pathbreaking work in the philosophy of science, is the author of such classic works as *The Science Question in Feminism, Whose Science? Whose Knowledge?* and *Is Science Multicultural? Postcolonialisms, Feminisms, and Epistemologies.* She is the author or editor of eleven books, and for five years coedited *Signs: A Journal of Women in Culture and Society.*

RACE AND GENDER IN SCIENCE STUDIES

Science and Social Inequality
  —*Sandra Harding*

The University of Illinois Press
is a founding member of the
Association of American University Presses.

Composed in 10.5/13 Adobe Minion
with Meta display
by Type One, LLC
for the University of Illinois Press
Manufactured by Thomson-Shore, Inc.

University of Illinois Press
1325 South Oak Street
Champaign, IL 61820-6903
www.press.uillinois.edu